Building Cost Planning for the Design Team

Building Cost Planning for the Design Team

Second edition

Jim Smith
Professor of Urban Development
Bond University, Queensland, Australia

and

David Jaggar
Emeritus Professor
Liverpool John Moores University,
Liverpool, UK

AMSTERDAM • BOSTON • HEIDELBERG • LONDON • NEW YORK • OXFORD
PARIS • SAN DIEGO • SAN FRANCISCO • SINGAPORE • SYDNEY • TOKYO

Butterworth-Heinemann is an imprint of Elsevier
Linacre House, Jordan Hill, Oxford OX2 8DP, UK
30 Corporate Drive, Suite 400, Burlington, MA 01803, USA

First published 1998
Second edition 2007

British Library Cataloguing in Publication Data
A catalogue record for this book is available from the British Library

Library of Congress Cataloguing in Publication Data
A catalogue record for this book is available from the Library of Congress

ISBN 13: 978 0 7506 8016 5
ISBN 10: 0 7506 8016 4

For information on all Newnes publications visit
our website at http://books.elsevier.com

Typeset by Charon Tec Ltd (A Macmillan Company), Chennai, India
www.charontec.com
Printed and bound in Great Britain

06 07 08 09 10   10 9 8 7 6 5 4 3 2 1

Dedication

Dedicated to all the fans at
Hillsborough and Anfield

Contents

Preface

The Tower

"We'd like to touch the stars", they cried, and after,
"We've got to touch the stars. But how?"
An able-brained bastard told them: "Build the Tower of Babel.
Start now, get moving. Dig holes, sink a shaft. A-
Rise, arouse, raise rafter after rafter,
Get bricks, sand, limestone, scaffolding and cable.
I'm clerk of works, fetch me a chair and table".
God meanwhile well nigh pissed himself with laughter.

GIUSEPPE GIOACHINO BELLI (1791–1863)
(QUOTED IN A. BURGESS, ABBA ABBA,
LITTLE BROWN & CO., BOSTON, USA, 1977, P. 102)

Belli's irreverent poem about the building of the Tower of Babel seems to capture the essence of building. It's an activity that is often frustrating and rewarding at the same time. However, the one enduring characteristic is that it produces *real* things that society can use to create more wealth and well-being. Whilst a great deal of criticism is levelled at our industry (and some of it is deserved) the central fact remains we do make a significant contribution to the economy by building, converting and renovating property of all kinds. Unfortunately the problems and disasters seem to receive all the attention in the media. However, a greater number of projects are well built, affordable and timely and they provide a source of pride for a long period of time to all the project participants, users and owners.

Cost planning may seem to be a more insignificant and ephemeral activity in the scheme of things. Nonetheless, it does make a small, but often meaningful contribution to the development and the successful completion of a project. Clients certainly value a professional cost control service and have been the driving force for change in the arrangements to procure buildings. In the turbulent construction environment clients have forced major structural and procedural change on the industry through the increased use of design and construct arrangements, integrated project management teams, novation, partnering, benchmarking, re-engineering, management contracting, design and build, private finance initiatives, public and private partnerships and many other developments. The old client maxim seems to be even more apposite: that is, after deciding to build clients want to know two things, *When can I have it? How much will it cost?*

In fact, it is surprising that it has taken so long for the supremacy of client needs to be fully recognised by many of the industry's professionals. The subdivision of services, developed as a result of history rather than as a response to client needs, seems to be changing rapidly. Clients naturally have no interest in the traditional division of the construction disciplines within a conventional procurement method where it is obviously not satisfying their needs. These client pressures have led to a major reorientation of design and construction services within our industry that will have a permanent and lasting impact on the way we service clients. Whilst design, project management, cost control, quality control, planning, competitive tendering and contractor selection are enduring activities in construction, the means of procuring and delivering them is many and varied and subject to rapid and continuing change. The modern construction adviser must be aware of these trends and be prepared to adapt in this volatile and demanding environment. Rapid change and greater demands for accountability and high quality building supported by rigorously applied professional services will be the standard operating construction climate for the foreseeable future. Under these demanding conditions cost planning must be seen as the responsibility of the whole design and construction team. It is not a specialised, compartmented, esoteric activity under the restricted responsibility of the cost planner/quantity surveyor/estimator on the team.

This book's central thesis defines cost planning as a team responsibility and it attempts to broaden the limited vision of conventional cost planning by considering such issues as value, use and effectiveness of the completed building. Such an approach places greater demands on the cost planner with a commensurate need for the design team to become more involved in the cost and value consequences of their decisions. These are new and exacting challenges being placed on the design team. However, it is imperative that we continue to provide high quality professional services that clients and the community desire and that are suited to their environment. However, certain requirements will remain constant and we must always strive to deliver well-designed, economical and effective buildings that are valued by their owners and users both now and in the future.

This is a research based and professional text concerned with economic environment of the building industry and the role of cost planning in its activities. It presents the range of knowledge and information that a construction professional needs when working as a financial adviser and member of the design and construction team on a project. It is based on extensive research and years of practical experience in the field. The primary aim of this book is to provide you with the background, fundamental skills and knowledge to operate as a building professional in the area of economic advice and decision-making at the project level.

The content of the book is divided into two parts:

- Context
- Cost planning the design stages

The context covers the environment in which cost planning and design activities take place. The issues considered play an important role in guiding project decisions: characteristics of the construction industry, the development process, design economics and design stage estimating. Knowledge of these topics enables the cost planner to make a more significant contribution to design team decision-making.

Cost planning the design stages is the heart of the book detailing the development of a project through the design stages from the early stages of *Establishing the Need* for the project to the brief and on to tender documentation. These four chapters describe the cost planning process in theory and practice, discussing the issues involved from a design team and client perspective. The descriptions of the early stages are based on the latest research by the authors.

In addition to the references, there are further reading items listed at the end of each chapter. These are required because no text today can stand alone as the sole source of information. These texts are valuable sources of further information and each has a contribution to make to this subject.

A significant feature of this text is the use of in-text questions and review questions. This range of questions has been included to provide you with the opportunity to reflect on the material and possibly apply it in your own work or studies. In many cases the questions are designed to provoke a new line of thought and possibly provide a different perspective on the material. You should answer these questions, compare your answer with the suggested answer provided and then review the text material if necessary, before proceeding to the next section or chapter. The suggested answers are not intended as the only answers to the questions posed, but are more in the nature of a discourse, and they are certainly not model solutions.

Another important attribute of this text is the reference and use of the Building Cost Information Service (BCIS) On-Line cost database. Sample data is provided throughout Part B, which describes the cost planning process. The authors believe that access to this type of cost data is indispensable to modern cost planning.

Whilst computers and access to cost databases is important this text is also concerned with fundamental principles, developing a thorough understanding of the process and achieving the financial aims of a project within a framework of quality, time and value. It is important that readers understand these timeless principles and their application and then use the ever changing and expanding computer software to extend and improve their application. The computers and the software will undoubtedly change, but the need to understand the fundamental concepts will not.

The authors' experience of cost planning in four countries (UK, Australia, Singapore and Hong Kong), also confirms the need to fully understand the basic principles of a subject. The context and the detail in which these activities are carried out may change, but the enduring principles that we both first learned in the UK have lasted almost unchanged over many years and in several locations. Many of the references in this book cross national boundaries and testify to the fact that the principles of cost planning are universal. Whilst this book has been written for UK conditions, it can be used by students and practitioners in other locations and jurisdictions.

Jim Smith
Melbourne, Australia
March 2006

David Jaggar
Liverpool, UK

Acknowledgements

The material in this book has been developed by the authors from research, teaching and professional experience over a long period of time. It is therefore difficult to identify all the individuals and organisations who have contributed to its development.

- Jim Georgiou, Lecturer and colleague, School of Architecture and Building, Deakin University. Jim's willing assistance and design and drafting skills have been used throughout the text.

- Alex Hollingsworth, Senior Commissioning Editor, Elsevier

- Lanh Te, Editorial Assistant, Building, Construction and Civil Engineering at Elsevier

- Joe Martin, Executive Director of the Building Cost Information Service (BCIS) Ltd, London, for permission to use extracts from the BCIS Online cost database and the Standard Form of Cost Analysis in the Appendix.

- Christine Morriss, Director, Rawlhouse Publishing Pty Ltd, Perth, Western Australia, for permission to use an extract of elemental costs from the *Australian Construction Handbook* (2006).

- Last but not least, our long-suffering and patient families – Edwina, Ben and Catherine and Christine. They have seen this book grow over a period of 10 years. They are relieved to know that whilst a book is never finished, a temporary halt has been called.

- To all those other people we have forgotten, our apologies.

Part A
CONTEXT

Chapter 1

Building and Construction Industry Characteristics

Cheapness in itself is no virtue: it is well worth to pay a little more, if as a result the gain in value exceeds the extra cost.

P. A. Stone (1980)

Chapter preview

Chapters 1 to 3 examine the development process, the construction industry and the design framework in which decisions are made in cost planning. These chapters provide the basic economic and organisational background for the cost planner. Knowledge of the economic and design environments are invaluable when working on a project with a client and design team. An appreciation of this setting will make the cost planner more aware of the effect of industry trends on his or her cost planning activities. This chapter is designed to make the cost planner aware of certain economic factors; however, it is not a comprehensive study of building economics. These studies are found in the more specialised texts noted in the Further Reading at the end of the chapter.

The cost planning in this text is focused on the building industry. It forms part of the broad discipline of building or construction economics. Building economics certainly requires an understanding of the cost planning techniques to practice cost control in the building design process, but more importantly, it is also necessary to appreciate the economic, industrial, environmental, social and project dynamics and environments in which development takes place. The techniques of cost planning can be readily understood and applied on a project. However, the financial success of the project is influenced by a multitude of economic and environmental factors, variables and decisions taken by many individuals and private and public bodies and organisations. These factors are often beyond the influence and control of the design team and the whole process can be described as complex. The Property Services Agency in the UK captures the essence of the situation accurately:

> All construction projects begin with the recognition of an opportunity or of a problem. To turn this recognition into a finished building or some other construction requires a multitude of decisions. These decisions concern amongst other things the location, size, quality, complexity, social and economic influence, time scale, organisation and cost of the project. They

are made by a multitude of people. Some of these decision-makers are
members of the body which originally recognised the opportunity or
confronted the problem, but many belong to other firms, practices and
authorities. It is a matter requiring much skill and knowledge to guide a
construction project through this complex decision-making process.

Property Services Agency (1981: 5)

We cannot list all the factors likely to influence the economic or financial
outcome of a project. However, we do attempt to broaden the traditional
focus of cost planning by placing it in a wider building economics context.
An appreciation of the development process, the participants in that process
and the role of the construction industry are the first steps in increasing one's
awareness of the complex environment in which cost planning activities
take place.

Financial control of building and construction at the design stages, within a
broad economic framework, is our focus of study in this introductory
chapter.

THE CONSTRUCTION INDUSTRY

The construction industry in any country is the sector of the economy that
plans, designs, constructs, alters, refurbishes, maintains, repairs and eventually demolishes buildings of all kinds. The industry also includes civil
engineering works and other similar structures, and their related mechanical
and electrical services. So, the term, construction industry is a generic expression for all sectors of the industry. As noted earlier our focus is on the *building
sector* of the construction industry. We shall describe all those parts that
contribute to output in the building industry and we shall not include the
civil engineering and infrastructure sectors. Similarly, some definitions of the
industry relate only to the contracting part of it, the segment that undertakes
the actual physical construction, others include materials producers and
plant manufacturers. The above description of the industry also includes the
specialist building professions (architects, quantity surveyors, project
managers, planners, engineers). It also comprises contractors who may work
as main, specialist or subcontractors in one or a number of fields of activity,
skilled tradesmen and unskilled workers.

The construction industry is a service industry that obtains its inputs from
other sectors of the economy, such as manufacturing, financial services, local
government and commercial sectors. For instance, it also involves the industrial
sectors supplying materials such as cement, bricks, tiles, plasterboard, timber
and other products used in buildings of all kinds. At the completion of the
building works the property sector provides the real estate and advertising
skills to market and sell or rent the buildings or spaces they provide. In
addition, both the public and private sectors are engaged in the industry as

clients and participants. When one considers the whole range of participants, authorities and range of activities covered in the regulation, design, construction and delivery of the products of the building industry it can be readily seen that it has great importance in the economy of a country or region and in the well-being of the communities that make up society.

Development and The Construction Industry

We need to distinguish between the cost and the value of a building. The final product (the building) has a cost (to the client) depending on its size, the quality of materials and services, the prevailing levels of prices for the inputs and the profit charged by the contractor. This is different from its value, which is related to the interaction between cost and worth, desirability, utility and quality. Value is a more nebulous factor as it varies with location and time. So, a building, new or existing, will change in value over time as this depends on market conditions, its location and the levels of supply and demand for such buildings. Some of the elements captured in cost and value are summarised in Figure 1.1.

Put another way, a client obviously aims to minimise the cost of a project whilst maximising its value. When building costs are lower (preferably much lower) than the value of the project then building activity will be strong. If cost and value become closer then building activity will reduce until the stage that no new works are commissioned. There have been periods in most countries where economic conditions (the market) have deteriorated to such an extent that the costs of building have exceeded the market value. In this

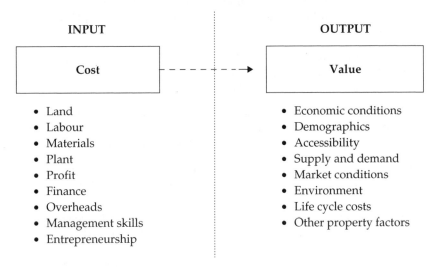

Figure 1.1 Costs (input) and Value (output) in the Building Industry

situation the construction industry, in common with the economy goes into recession until values of new building exceed the cost of construction. Note, that this is a simplistic analysis with static terms of reference. The actual situation in the industry, the building market and the property market, is much more complex and is often difficult to evaluate, and certainly difficult to predict its future activity levels.

Brandon (1984) calls value, or return, the *missing dimension* and makes a cogent argument for the design team to take it into account in all their decisions. Cost and value in construction are the two primary issues in the development environment and the cost planning process. Whilst the emphasis in this text is on the cost of construction, comment will be made on the value implications of certain decisions and for the development process to succeed a balance must be struck between these two important concepts. For a discussion of some of the issues involved, see Latham (1994), Keel and Douglas (1994), Atkin and Flanagan (1995) and Egan (1998).

Key Terms

It is necessary to identify some of the key terms used when discussing and analysing the industry and of their significance in discussing the statistical characteristics of this sector of the economy.

Some Terms Defined

- *Cost*: The *internal* amount paid for materials, goods and services including labour by the supplier or contractor. In terms of construction, this amount is often termed the capital cost or the initial cost and includes all the items on the cost side of the equation in Figure 1.1.

- *Price*: This is the *external* amount that needs to be paid for or charged for materials, goods and services including the vendor's (or contractor's) cost, profit and other charges.

- *Value*: The open market price for an asset, goods and services at a particular time. As noted earlier this is a difficult concept to define in advance of the market determining the specific value.

- *Worth*: The individual's subjective perception of the value for goods and services. In terms of design and construction it is the quality that renders something desirable, useful or valuable.

Hillebrandt (2000: 16) has a fascinating discussion of price and cost:

It is sometimes easy to confuse price and cost. That is, because what is a price to one person may be a cost to someone else. A cement producer will sell cement to a contractor at a figure which is the price to him. It is a price to the contractors too, in the sense that it represents the exchange terms dictated by the market. The contractor, however, will view what he had to pay for the cement as a cost, because it is the cost of one of the inputs to his construction projects.

It is interesting to note that if we followed the definition of the terms for cost and price above our studies in cost planning should be more accurately described as building price planning, as we are attempting to predict or forecast the price charged to clients by the market and building contractors. We still need to understand how building costs are derived and calculated, but it is the building prices to clients that we are mainly interested in. However, since the practice of, and term, cost planning is well established we are going to retain the expression, cost planning throughout this text.

Cost and Value in Practice

Broadbent (1982) attempts to grapple with some of the issues of cost and value. In particular, he analyses the approach of John Portman, an American architect who put his beliefs on the value of architectural quality to the test in his own self-developed project. It is useful to quote Broadbent (1982) at length because what he describes captures the essence of what we mean by value and it goes to the heart of design team operations:

> I hope it is clear by now that I am concerned with *real* economics, which puts the value of things above the price rather than that spurious kind which equates cheapness with *economy*. I don't even have to speculate as to how the economic value of quality in architecture may be demonstrated. For the fact it has been demonstrated, by John Portman, of Atlanta, Georgia. Portman has said on more than one occasion that as an architect at first he felt constrained. Realtors (estate agents), cost accoun(tan)ts, and above all, clients each naturally had views as to what kinds of buildings were wanted, and how much each building should cost.
>
> Portman felt this unduly restricting, so having read the fine print of the American Institute of Architects' Code of Conduct, he decided to become his own developer, or … to think of real estate architecturally, and architecture entrepreneurially …
>
> Portman felt that if he offered the right package of architectural quality and real estate potential, then the sites, and the necessary finance, would come his way, which they did!
>
> … the Regency hotel presented Portman with his first opportunity to use architectural quality as an essential component of the economic equation.
>
> Originally, Portman, had designed a fairly *normal* hotel with a slab containing several storeys of corridors with bedrooms down either side,

supported over a *podium* containing the lobby, public rooms and so on. But then, as he said:

'I didn't want the hotel to be just another set of bedrooms … a cramped thing with … a dull and dreary lobby … elevators over in the corner … a hotel room with a bed, a chair and a hole in the outside wall'.

Instead of that: 'I wanted to explode the hotel: to open it up, to create a grandeur of space, almost a resort, in the center of the city'.

And so he did, with the famous 22-storey atrium, a vast, rectangular open space with glass elevators gliding up the face of a concrete wall to a roof top restaurant and open balcony access to the bedrooms.

Portman showed his designs to Conrad Hilton, whose considered comment was: 'That concrete monster will never fly'. The Sheraton, Loews and Western Hotel chains were equally pessimistic for they knew intimately, the economics of running hotels. They really did not see how Portman's could be viable. For one thing it was strange, which they felt, might deter potential customers and, for another, they could not see any possibility of financial return on that vast, unusable atrium space.

Of course, it was an instant success and within three months demand for rooms was such (94.6% occupancy) that Portman had to design an extension.

The rest is history as they say! It is difficult to imagine that modern hotels, some offices and shopping centres would not use the same atrium principle in their layout. Who would have thought that a design concept which had lain fallow since Roman times have such a stunning impact on modern development projects? However, such a unique approach requires an intimate knowledge of the market, a great deal of creativity, but more importantly, a good dose of courage to pursue the idea using one's own financial resources and unerring confidence in one's own judgement.

However, for this one example of success there are many more examples of building projects that failed to take account of market gained through prudent analysis and research of supply, demand and demographics of the specific market.

Question 1.1

Broadbent's discussion of cost and value in the extract above (despite being written in 1982) still has implications for the way design team members advise and interact with each other and with the client.

After reading the quoted material above, what does it say about design team advice, cost and value of projects?

Measuring Output

The construction industry produces, in each year, a wide variety of buildings, new and renovated facilities and works and, within each region, the mix of types of completed works varies over time. Certain items, for example, an airport, a power station, harbour or a dam may not be constructed for decades. When such facilities are being constructed it may unduly influence the value of output in that region whilst that work continues. Thus, it is difficult to measure and compare levels of construction activity objectively over time. Measuring buildings by floor area, roads by length (and widths) and bridges by number (and spans), sewers and drains by length (and internal dimensions) would result in a confusing array of data even without considering the need for stating various types of construction, finishes, services and so on. To simplify the comparison, monetary costs or more accurately, using our earlier definitions, *prices* (to clients) of all items of construction are commonly used. This is not without its problems. Costs change with time owing to inflation (which is mostly offset by applying a common time for measuring costs and variations in the industry's efficiency). Secondly, soil conditions, topography and climate affect the cost of construction in different parts of the country and internationally. Thirdly, as noted above a few mega-projects of high cost may distort the picture in a region for some time.

INFORMATION ON THE BUILDING AND CONSTRUCTION INDUSTRY

Data on the construction industry are usually collected by government and typically are published quarterly and annually by government bodies. In the UK, it is the Central Statistical Office and in Australia, this is the Australian Bureau of Statistics (ABS). Treatment of the construction industry as a major industrial sector of the national economy means that it is also frequently mentioned in treasury reports, economic surveys and by government departments, particularly those involved in housing, education, employment and public works. Newspapers, construction and housing industry associations, a number of quantity surveying and cost consultancies, and many observers and participants in the industry also publish summaries of trends of these statistical data. The Royal Institution of Chartered Surveyors (RICS) through the Building Cost Information Service (BCIS) publishes a comprehensive analysis of all construction statistics for use by its membership and the wider community. These centrally collected construction statistics are essential for future planning, predicting industry workload and capacity, and are used in influencing government and private sector decisions on whether, and when, to proceed with certain projects to gain the best value for the investment they are making.

Again it is necessary to consider some of the common expressions used in these publications used to describe the activities and output of the construction industry.

Gross Output

The gross output of the building and construction industry is the total value of work carried out on both completed and uncompleted projects. The statistics also provide data on work commissioned or planned (*on the drawing board*, so to speak) to provide advance notice of the impending commencement of construction works in a region.

Capacity

The capacity of the construction industry is a measure of its optimum possible (not necessarily the maximum) output per annum. The maximum output for the industry may result in shortages of resources and bidding for the scarce ones that tends to have inflationary effects without an increase in productivity. The costs and the prices charged for the products of the industry would necessarily rise. The optimum in terms of the industry's capacity is a more appropriate goal to aim for. For the economist, this is the point at which the industry is utilising all its inputs to the best possible extent. The aim of the industry and the Government is to attempt to use the resources of the industry in the most efficient and effective manner. That is, they strive to prevent 'overheating', which raise prices without a rise in output.

Resources can be either under-utilised owing to poor organisation and management, or overstretched in a boom period, resulting in 'inflationary overheating.' Neither situation is desirable. In the medium and long term, of course, improving efficiency and productivity in both design and construction can enhance capacity.

Construction Cycle

During a period of expanding economic activity and good future prospects, incomes rise generally and individuals can purchase houses and consumer goods, companies need larger or improved premises to meet increased demand for their goods and services, and government may want to improve on the social infrastructure (e.g. housing, hospitals), so as to stimulate further growth (roads, communications networks) or avoid shortages of trained personnel (schools, training centres). Demand for construction increases. As the industry uses up the previously under-utilised capacity more skilled personnel are attracted from allied industries and construction firms are able to pay more. The industry becomes generally inefficient with the influx of inexperienced enterprises and workers. Entry barriers to new enterprises may be established by the authorities by the use, for example, of minimum capital requirements, qualifications and registration or licensing requirements and the like. Similarly, skill and qualification controls may be applied to employees such as electricians or plumbers to ensure minimum levels of competence. Materials may also become in short supply and prices rise. Government may step in to control the inflationary spiral through interest rate rises, but a point is reached when demand begins to fall either because construction items are considered too expensive or the prospects of the economy become less bright.

In a period of low demand, tender prices are lower as competition is keener and less efficient or uncommitted firms are forced out of business. Unemployment rises and investment in training and equipment is curtailed. Materials producers and plant manufacturers cut back on production and shelve plans for expansion. The capacity of the industry is reduced. Though most economic activities are subject to business cycles, the ease with which investment in construction can be postponed makes the difference between the maximum and minimum demand greater than that for most other activities. The greater the amplitude of the fluctuations and their frequency the less able is the industry to meet future increases in demand as it cannot plan for the future with confidence.

This situation has been summarised in the real estate market by the following widely known and frequently published diagram (anonymous author) in the mass circulation press (see Figure 1.2).

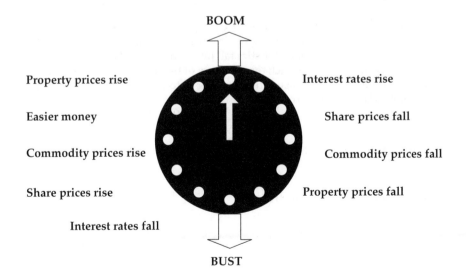

Figure 1.2 The Investment Clock
Source: Unknown.

Although Figure 1.2 describes the process as an investment 'clock', it is more like a cycle of activities and characteristics. The sequence of events in most situations will most likely follow that given in the figure. It represents an idealised version of reality, but it is also useful in understanding the investment environment. It should also be noted that the period of time taken with each event is not likely to be as equal as the 'clock' implies. The length of any distinct period will depend on the circumstances; some may be extended, others may be compressed and even merged due to the economic conditions and possible government intervention.

The significance of this 'clock' to the construction industry is that investment in property is likely to have an effect on the demand for construction. When interest rates fall, not only will it encourage investors in commodities and shares, but it will also encourage greater economic activity and demand for services and products. In some cases this will revitalise greater investment in buildings and plant to expand the productive capacity of the economy. Similarly, the cost of borrowed money for building projects will be lower and this is likely to encourage more investment in construction. Knowledge of the sequence of events in the investment 'clock' by the design team and members of the construction industry will make them more aware of the present and likely changes in their future workload and activities due to changes in the economic patterns. It could also alert them to opportunities and risks in markets and regions with a differently synchronised investment clock and economy.

CHARACTERISTICS OF THE CONSTRUCTION INDUSTRY

> ... construction has characteristics which, separately, are shared by other industries, but in combination appear in construction alone, justifying its industry status. These characteristics fall into four main groups: the physical nature of the product; the structure of the industry, together with the organisation of the construction process; determinants of demand; and the method of price determination.
>
> P. M. Hillebrandt (2000: 4–5)

We shall analyse the points summarised in this quotation.

Features of the Construction Industry

As noted above by Hillebrandt (2000), we contend that the construction industry has many features that are found individually in other industries. However, these features combine and interact with each other in such a way as to make the construction industry difficult to plan for, forecast, manage and control.

These features include:

- Size.

- Government as a primary client.

- High cost of product.

- Response to demand and structure of industry.

- Time delays.

- Organisation and participants.

- Impact of technology.

- Problems such as slow decision-making, misunderstandings and conflict between the various parties, poor management, illegal and unethical activities (For example, in the UK see the Latham (1994) Report and the Egan (1998) Report and in Australia, see Gyles et al., New South Wales Royal Commission (1992) – The Gyles Report and the Cole (2003) Report, Final report of the Royal Commission into the Building and Construction Industry, Volumes 1 to 9.).

These features and how they interact will now be discussed.

Size of Industry in National Economy

All governments recognise the important role that the construction industry plays in the economy of every state and all nations.

Why is this?

- The construction industry typically and consistently contributes a sizeable proportion of the gross national product – between 7% and 10%. This percentage would be larger if all the associated industries and financial, real estate and facilities, and property management activities were included in the construction sector.

- It employs a significant number of the working population – generally between 6% and 8%.

- It is responsible for more than half of the national capital formation.

- Its products form the factors of production (land and buildings) for almost all the other industrial sectors.

- Shelter, living accommodation and transportation are considered the basic necessities of modern life. These are provided by the construction industry.

Socio-economic development is concerned with expanding the productive capacity of the national economy to increase or improve the quality and extent of goods and services available to the community. That is, improve standards of living and economic well-being. Thus, the level of construction activity is often used as a measure for socio-economic development and progress within a society. The construction industry has an indispensable role since it contributes so much to capital formation and housing. It has been detected that acceptance of long periods of decline and neglect of the construction industry has an unfavourable effect on future economic and social development. For instance, neglect of infrastructure provision and its maintenance and improvement can have a marked effect on transportation costs and efficiencies, which will eventually influence the rate of economic growth and overall international competitiveness.

Due to its significance and size, in most economies, governments cannot ignore the industry. In many cases governments have developed policies and established advisory bodies (such as the Construction Industry Development Agency (CIDA) in Australia (from 1990 to 1995) and the Australian Construction Industry Forum (ACIF) and the Australian Procurement and Construction Council (APCC)) to encourage construction efficiency and improvement. In the UK and Singapore and other countries, similar bodies have been established (the Construction Development Board in Singapore and in the UK, the Construction Industry Development Board and Construction Clients' Forum). The industry is also intimately connected with many other sectors of the economy in a complex network of inputs such as materials and components supplies and with outputs in the form of its product, buildings and facilities. Consequently, a wide range of government measures such as taxation changes, import controls/incentives, investment shifts and industrial relations policies; all have ramifications for the industry's future workload and level of activity.

Government as a Major Client

In the past, governments commonly finance infrastructure projects in the civil engineering sector of the industry, such as roads, bridges, tunnels, freeways and drainage projects. Whilst governments are still keenly interested in the provision of these facilities, the form of finance and ownership of such facilities has undergone a dramatic change. In the UK and Australia, the most common method of procuring these facilities is by using private sector investment money and expertise through private finance initiatives (PFIs) and public and private partnerships (PPPs) in the UK. Similar terms are used in Australia. Government is also the dominant client for social projects such as hospitals, schools, police stations, courts, prisons and similar non-profit making schemes. Some of these facilities have also been the subject of PFIs and PPPs in both countries. Other investors in infrastructure include the major utility boards and corporations (water, gas, electricity) whose major source of funding is from the consumer and share market investors. Time will tell whether these forms of financing and procurement methods will continue into the future.

The level of public sector investment in construction at any time is related to the government's concern with a complex set of factors:

- controlling inflation;
- attempting to secure a reasonable level of employment;
- balancing the budget;
- ensuring a healthy balance of payments;
- achieving an increase in certain critical facilities for a specific event (sports events such as the Olympics) or to solve a supply crisis in a sector (or specific) region such as housing.

Some writers believe that governments should use their power as a major client and manipulate their development budgets to regulate or manage the economy. This is fraught with danger. In practice, the results have rarely coincided with the aims as expressed in policy statements. Commonly, the government expects a much more rapid response to an injection of funding than actually eventuates. Planning delays, inter-government and inter-departmental conflicts, all contribute to delays before the construction industry can respond. Government intervention also 'contaminates' market forces and may dilute the strength of the market as a regulator and promoter of activity.

Question 1.2
Building Construction Statistics

Check the sources of building and construction statistics in your country or region of interest.

- What was the total value of the output of the industry for new construction, repairs and maintenance for the past 3 years?
- What was the value of public sector construction compared to private sector activity?

After looking through some of the official publications for construction statistics briefly consider:

- What are their main contents?
- What do they tell us about the construction industry?
- Are they useful in our work as design team members and client advisers?

High Cost of Product

The outputs of the building industry are very costly. The majority of clients in the private sector have to rely on borrowed funds to secure their purchase. Consequently, the effective price of a building is not purely its capital cost, but it also includes the cost of servicing the loan or loans. As a result, the important factors to customers of the industry are:

- availability of funds for lending;
- rate of interest, or cost of lending;
- duration of the loan;
- inflation, now and expected in the future.

In summary, the Central Bank's monetary decisions and the government's fiscal and monetary policies will influence interest rates, which in turn can have an effect on the demand for construction.

Response to Demand and Structure of the Industry

Demand usually consists of a building works for a specific client. The order or contract for the works commonly has the following features:

- *Cost*: A (largely) predetermined cost (tender).
- *Time*: Undertaken within an agreed or defined period (project duration).
- *Location*: A fixed location (the site).
- *Documentation*: Built according to client-supplied documents (such as traditional specifications, drawings, schedules and possibly, bills of quantities) or provided by the successful tenderer depending on the method of procurement.

Tendering methods may vary and the risk transferred to each party under the various contract arrangements makes construction a very uncertain business (Cartlidge, 2003). Uncertainties (in information, site conditions, interest rates, project content, weather, materials prices, lack of specialised skills and industrial relations) can combine to lead to big losses on projects (Edwards and Bowen, 2004). The main contractor often has little control or influence over the risks that have to be assumed on many projects. Insurance for many of these risks is not available.

The nature of construction projects: fixed site locations; bulky materials and components; varying project sizes and types; and specific combinations of resources – materials and labour, plant, means that resources cannot be easily substituted between projects. Any strategy for influencing demand must take account of these characteristics.

In response to this uncertainty in demand and project conditions, the construction industry relies substantially on the subcontract system to complete its projects. The role of the contractor is often to manage the subcontractors carrying out the actual work. Unfortunately, many of these subcontract firms have insufficient capital for investment in plant and equipment, and may even have a limited commitment to the industry. Training of the skilled personnel required on most modern projects is neglected and many firms consider this investment in labour to be wasted as the newly trained operative may be 'poached' by rivals.

Typically, in most countries, there are only a few large and highly capitalised contractors operating on a nationwide, and sometimes on an international basis. Below the large firms are a number of medium-sized firms, and below this a multitude of small contractors and subcontractors.

In most countries a consistent feature of the building industry is that a large proportion of the total firms are small firms with 20 employees or less. For example, in Australia, these small firms represent over two-thirds of the total number of firms. The same situation applies in most countries. The construction industry is dominated by these small businesses and this must be recognised by people within the industry and by decision-makers who may wish to influence its characteristics. Large construction companies with more than 100 employees may enjoy a higher profile with the public through the projects they are involved in, but the reality is that they are only a minor proportion of the total industry in terms of the total number of people employed.

Time Delays

The design process passes through various stages, which will be discussed in detail in later chapters. Using the Royal Institute of British Architects (RIBA) Plan of Work (RIBA, 1998, 2000) as the model the six major stages are appraisal, strategic briefing, outline proposals, detailed proposals, final proposals and production information. These stages can be managed once the major project decisions have been made. However, the pre-design stages under the control and influence of the client, authorities and other stakeholders can be more problematical and subject to excessive time delays.

Each stage can be complex and often takes a considerable amount of time. Many major projects can take from 3 to 5 years (or more) to complete from the decision to build to handover of the final building. Even the simplest building project, following a familiar process, can take from 9 to 12 months to complete. This pre-design process has been documented by one of the few authors in this field (Woodhead, 1999, 2000).

The lack of predictability in the time factor increases the risk to both the client, who may occupy or lease the completed building, and the contractor who tenders in advance of the actual work. The decision to proceed with a project may be made in a totally different economic and market environment from the one in which the project is completed. It may take 1, 2 or 3 or more years (possibly 5 years according to Woodhead, 1999) to complete a project and the market situation can change from highly favourable (boom) to depressing (bust). Design decisions are also made well in advance of the building's occupation and the client and design groups carry an onerous responsibility for predicting the needs of the users and their clients on immediate occupation and throughout the life of the building. This adds to the risk of developments and has contributed either to the massive success or the sickening failure of building projects and contributed to the insolvency of many development companies.

Organisation and Participants

Construction projects involve a complex network of participants and stakeholders, each with specialised expertise and diverse responsibilities. In general, the larger the project the greater is the number of diverse skills that are necessary. Despite the fact that rights, duties and obligations are described in the contract documents, formal and informal responsibilities can exist, which differ from one project to another, and also at different stages of the same project.

Tasks often overlap and are interdependent and performed in a dynamic, complex environment subject to changing pressures. A rigid set of rules and procedures can assist in the simple and straightforward projects, but they also tend to hinder the suitable harnessing of the skills and initiative of the participants. Delays, stoppages, conflict, misunderstandings and abortive work are too common within the industry.

The ongoing efficiency of the industry is certainly not improved by the traditional approach of many construction projects, which encourage the disintegration of project teams once the project is completed. Project arrangements and procurement methods such as Design and Build, Project Management, Build-Own-Operate-Transfer (BOOT) and Construction Management offer prospects of improved relationships and fewer problems (Walker and Hampson, 2003). These approaches are likely to increase in the future, with possible positive effects on the time, cost and quality of the final product.

Some commentators (Latham, 1995; Egan, 1998; Gray and Hughes, 2001; Jaggar et al., 2002; Walker, 2002) have noted that the separation of design and construction into two teams, often in conflict and often relegating the requirements of the client to a secondary or even tertiary level of priorities, may be a basic defect in the industry's traditional organisational arrangements. It is interesting to note that many clients in recent years have favoured design–construct contracts and project management arrangements where the traditional design construct divide is brought close together or bridged. Certainly, a major aim of these relationship-based procurement systems is to reduce conflict and areas for dispute.

Impact of Technology

The construction industry draws from an extensive range of products and technologies. New techniques and methods may replace the old. However, more commonly, new ones are added to the existing range. The technological spectrum often extends from the simple craft-based labour-intensive activities to the most advanced, including prefabricated and industrialised buildings.

This range can bring in its own problems to each project as the industry can make a choice from many feasible combinations of different materials, techniques and systems. The constraints on choice may be imposed by several factors: available resources; weather conditions; designer's opinions; location; skills available; building and planning regulations; occupational health and safety.

The nature of design and construction activity makes its complete industrialisation difficult, possibly unrealisable. The industry is more labour-intensive than most others. This makes it easy for new firms to enter and to increase the competition within it. Unfortunately, this easy entry makes it difficult to rationalise, which hinders any progress towards greater efficiency.

Problems

Unfortunately, the client's view of the industry is not a positive one. The industry suffers from a poor reputation and unfortunately problems in the industry gain wide publicity and reinforce the stereotype of poor performance. For instance, in the UK, the Latham Report (1994) and Egan Report (1998) and in Australia the Gyles Report (1992) and the Cole Report (2003) provided a depressing view of the performance of the construction industry. Everyone involved in the industry must be made aware of these perceptions and attempt to improve its efficiency, productivity, innovative capacity, legal and ethical compliance (probity) and customer relations or marketing. Good organisation and procedures for cost control and cost planning has a modest, if unglamorous role to play in improving the quality of decision-making, financial and time performance and forecasting within the industry.

The following is a selection of views from a 1993 report by the Construction Skills Training (Victoria in Australia) where it described some of the characteristics of our industry: a similar list could be drawn up for most construction industries in any country and Morton (2001) also provides a cogent summary of the UK construction industry:

- The construction industry has several sectors, is highly fragmented, makes extensive use of subcontractors.
- Difficult to establish objective benchmarks of performance.
- Subject to cyclical fluctuations in volumes of work.
- High degree of labour mobility.
- Little investment is made in an employee's skill or training.
- Contractual relations are often adversarial, conflict-based and destructive.
- Unrealistically low-bid tendering stimulates competition on cost alone.
- Commercial behaviour is dominated by the need to maintain cash flow, which puts financial pressure on suppliers, subcontractors and client.

Other problems include:

- Contractors lack vertical integration.
- Inadequate research into the construction industry's problems.
- Excessive reliance on credits and retention moneys.
- Numerous small firms serving a local and specialised market.
- Planning procedures for major projects that are often uncertain, causing delays and frustration.
- Low profit margins.
- Lack of cohesion between project team members.

All these problems can combine to cause major project disasters that confirm the community's worst anxieties about the industry. Although some of the problems listed above can be eradicated, or minimised by better internal arrangements within the industry, it is also apparent that the industry's structure and practices have evolved in response to the nature of its dynamic working environment.

Many writers and bodies have been critical of the industry and point out that to improve the industry's performance, the firms, participants and decision-makers, must increase their internal productivity and efficiency. At the same time, the industry must be more flexible and adaptable to improve its ability to cope with a continually changing operating environment. For an interesting review of the various reports into the construction industry in the UK since 1944 to the present Murray and Langford (2003) have compiled a thoughtful summary of 12 national reports from that date. A similar review of construction reports was made in Australia by the Royal Commission into the Building and Construction Industry (2003). Both publications make very interesting reading with a sad lack of progress for the industry in each country.

Chapter Review

Cost management does not take place in a vacuum. Cost planning has evolved in the environment of traditional procurement methods. However, its principles are sufficiently robust and rigorous to adapt to other procurement environments such as design and build or other integrated forms of procurement. Cost management is part of the total development process and it is essential that the cost planner understand the context in which cost planning activities take place. The development process takes a great deal of time and involves many interacting participants and agencies. While the influence and effects of some of these individuals, groups and activities may appear remote, their impact may be critical to the successful outcome of a project. It is, therefore, necessary for everyone involved in the development process to understand fully its workings so that their activities and contribution can be improved by such knowledge. For cost planners, this awareness of the process should enable them to concentrate their activities and skills in those areas, which will have the greatest impact on the final outcome in terms of cost, quality and time.

An understanding of the construction industry and the building sector within that industry must be developed if cost planners are to operate to full effect. It is an industry that has a bad reputation arising out of poor management, lack of coordination and a diffuse labour force based on a poorly integrated subcontract system. These characteristics impinge not only on the direct costs of construction through the tendering system, but also on the indirect costs borne by the contracting organisations and the client in the final costs and price of the completed project.

The characteristics of the industry must be recognised by all participants in the development process. Only by recognising these, can progress be made towards improving the efficiency of the industry and the lowering of the construction costs levied on clients. Participants must recognise that by speeding up the time and quality of decision-making, construction costs can be significantly lowered. Thereby, clients will gain a better product for a lower price and this will lead to greater client satisfaction.

To repeat what was said earlier, many writers and bodies have been critical of the industry and point out that to improve the industry's performance, the firms, participants and decision-makers, must increase their internal productivity and efficiency. At the same time, the industry must be more flexible and adaptable to improve its ability to cope with a continually changing operating environment.

The project management concept of time/cost/quality is important to recognise in all our development activities at all levels and particularly in the cost planning domain. There is a tendency to view problems in one dimension, particularly where costs are being forecast, planned and controlled. It is imperative that a project must always be considered holistically and whilst cost minimisation may be a worthwhile and realistic objective it must always be borne in mind that cost savings could have ramifications on the time and quality factors of the project. That is, the value of the project. A cost objective should never be viewed in isolation and the cost planner must always remind her/himself of this fact.

This chapter focuses on the development process and the building industry. That is, it concentrates upon the context in which cost planning takes place. As a cost planner you must not only be aware, but also familiar with the process, the participants and the environment in which your activities are taking place. In this way you will become a more effective contributor to the total cost planning function.

Economic factors figure prominently in our everyday affairs. They are even more important in the development of a building project. Success or failure, continuation or termination is crucially dependent on the economic factors surrounding a project.

An understanding of the size, nature and characteristics of the building industry is an essential prerequisite for operating effectively within it. While some of the problems of the construction industry are emphasised in this chapter and may appear to be exaggerated by the community at large, the participants in the industry must not be complacent about them and be prepared to face up to them and rectify them wherever possible.

Concepts of cost, value and price were introduced, and while they may appear to be obvious, or even academic or trivial, it is essential for the cost planner/building economist/cost consultant to be aware of the distinctions when providing advice to clients and members of the design team. The notion of value is stressed in this chapter and must always be borne in mind throughout the life of a project. All clients want better quality and more *valuable* projects. Adding value is one of the key functions of the design team and should be a central pillar of all building design cost planning and design teamwork.

Finally, the sources of information for the construction industry must be identified. While these are fleeting snapshots of the industry at any particular time, they must be monitored regularly for the data they provide to assist the cost planning function. This review and inspection of statistical data must remain a feature of your continuing professional life and personal development activities.

Review Questions

What are perceived as the inherent weaknesses of the building industry?

Can they be remedied?

References

Atkin, B. and Flanagan, R. (1995) *Improving Value for Money in Construction: Guidance for Chartered Surveyors and their Clients*, Royal Institution of Chartered Surveyors, London, UK.

Brandon, P. S. (1984) Cost versus quality: a zero sum game, *Construction Management and Economics*, **2**, pp. 111–126.

Broadbent, G. (1982) *Design Economics and Quality* in *Building Cost Techniques: New Directions*, Brandon, P. S. (editor), E. & F. N. Spon, London, UK, pp. 41–60.

Cartlidge, D. (2003) *Procurement of Built Assets*, Architectural Press, Oxford.

Edwards, P. and Bowen, P. (2004) *Risk Management in Project Organisations*, Architectural Press, Oxford.

Egan, J. (1998) *Rethinking Construction*, Report from Construction Task Force, Department of the Environment, Transport and the Regions, London, UK.

Gray, C. and Hughes, W. (2001) *Building Design Management*, Butterworth-Heinemann, Oxford.

Gyles, R. V., Yeldham, D. A. and Holland, K. J. (1992) *Report of Hearings, Royal Commission into Productivity in the Building Industry in NSW* (Gyles Report),

Government of NSW, Sydney. (The Report consists of ten volumes: Vol. 1 – Corrupt Practices; Vol. 2 – Collusive Tendering; Vol. 3 – Report of Hearings; Vol. 4 – Illegal Activities; Vols. 5, 6 – Industrial Relations; Vol. 7 – Final Report; Vols. 8, 9, 10 – Appendices, Policy and Research Papers, Results.)

Hillebrandt, P. M. (2000) *Economic Theory and the Construction Industry*, 3rd edition, Macmillan Press, Basingstoke, UK.

Jaggar, D., Ross, A., Smith, J. and Love, P. (2002) *Building Design Cost Management*, Blackwell Publishing, Oxford.

Keel, D. and Douglas, I. (1994) *Client's Value Systems: A Scoping Study*, The Royal Institution of Chartered Surveyors, London, UK.

Latham, M. (1994) *Constructing the Team: Final Report of the Government/ Industry Review of Procurement and Contractual Arrangements in the UK Construction Industry*, HMSO, London, UK.

Morton, R. (2001) *Construction UK: Introduction to the Industry*, Blackwell Science, Oxford.

Murray, M. and Langford, D. (2003) *Construction Reports 1944–98*, Blackwell Science, Oxford.

Property Services Agency (1981) *Cost Planning and Computers*, Department of the Environment, London, UK.

RIBA (1998) *Outline Plan of Work 1998*, RIBA Publications, London, UK.

RIBA (2000) *The Architect's Plan of Work*, RIBA Publications, London, UK.

Royal Commission into the Building and Construction Industry (2003) *Summary of Findings and Recommendations (volume 1), National Perspective: Part 1 (volume 2), National Perspective: Part 2 (volume 3)*, Commonwealth of Australia, Canberra, Australia.

> *Note*: The findings of the Royal Commission are commonly referred to as the *Cole Report* with the title coming from the Royal Commissioner heading up the Inquiry, The Honourable Terence R. H. Cole. The report was published in 22 volumes. However, the essence of the construction industry and the Commission's findings are covered in the three volumes noted above.

Walker, A. (2002) *Project Management in Construction*, 4th edition, Blackwell Science, Oxford.

Walker, D. and Hampson, K. (2003) *Procurement Strategies: A Relationship-Based Approach*, Blackwell Publishing, Oxford.

Woodhead, R. M. (1999) *The Influence of Paradigms and Perspectives on the Decision to Build Undertaken by Large Experienced Clients of the UK Construction Industry*, Unpublished PhD Thesis, School of Civil Engineering, University of Leeds, UK.

Woodhead, R. M. (2000) Investigation of the early stages of project formulation, *Facilities*, **18** (13/14), pp. 524–534.

Further Reading

The following publications provide more detailed description and analysis of some of the economic issues presented in this topic:

Ashworth, A. (1999) *Cost Studies of Buildings*, 3rd edition, Longman Scientific and Technical, Harlow, UK.
    Chapter 1: Introduction, pp. 1–13.
    Chapter 2: The construction industry, pp. 14–22.

Bon, R. (1989) *Building as an Economic Process*, Prentice Hall, Englewood Cliffs, New Jersey.

Briscoe, G. (1988) *The Economics of the Construction Industry*, The Mitchell Publishing Co., London, UK.

Ferry, D. J. and Brandon, P. S. (1999) *Cost Planning of Buildings*, 7th edition, Blackwell Science, Oxford.
    Chapter 1: Introduction, pp. 3–7.

Hillebrandt, P. M. (2000) *Economic Theory and the Construction Industry*, 3rd edition, Macmillan Press, Basingstoke, UK.

Morton, R. and Jagger, D. (1995) *Design and the Economics of Building*, E & FN Spon, London, UK.
    Chapter 1: The economic significance of building design, pp. 3–9.
    Chapter 5: Human resources for building, pp. 79–107.
    Chapter 6: Technology: from steampower to robots, pp. 109–129.
    Chapter 7: Construction – A unique industry, pp. 131–153.

Raftery, J. (1991) *Principles of Building Economics*, BSP Professional Books, Oxford.

Ratcliffe, J. (1978) *An Introduction to Urban Land Administration*, The Estates Gazette Ltd., London, UK.

Runeson, G. (2000) *Building Economics*, UNSW Press, Sydney, Australia.

Stone, P. A. (1976) *Building Economy*, 2nd edition, Pergamon Press, Oxford.

Warren, M. (1993) *Economics for the Built Environment*, Butterworth-Heinemann, Oxford.
    Chapter 8: The construction industry, pp. 122–138.

SUGGESTED ANSWERS

Question 1.1: Cost, Value and Design

This quote is a thought provoking and challenging piece. It may still leave you confused and perplexed as to where you should apply your cost planning skills.

The aim of selecting this first quote is to attempt to break free of the stereotype of the quantity surveyor or cost planner as merely a cost accountant and to focus attention at the beginning of this text on the really important project issues: clients; the market; users and value or more importantly, adding extra value through more effective design team activities.

In any project activity the same questions should always be asked, 'What does the building actually do?' 'What is the value adding of my activities?' The design team members must always bear this in mind in their decision-making and if a feature or space does not contribute to the purpose of the building then one may question the validity of its inclusion. This means participants in the design team may have to cross over the traditional professional boundaries. The cost planner, or quantity surveyor, typically focuses on the capital costs. The broader economic evaluation is often ignored, although the newer techniques of life cycle costing and value management (discussed in Chapter 10) have broadened the base of the economic evaluation. These evaluations are ones that include the land value of the site and its surrounds, running costs and costs of cleaning, maintenance repairs and rehabilitation. The focus of this book is the capital costs, but the wider evaluation to include the comprehensive range of costs noted above *must* also be carried out.

The message should be clear; get involved in the whole development process, especially the market surveys, the briefing stage and feasibility studies. Do not hide your expertise behind a label of being the cost planner of the building costs; important though they are! Attempt to balance cost, time and quality for the benefit of the project and its users. Some believe that the architect is the only person who can strike the balance; it requires the efforts of everyone in the design team. Good advice is essential to the correct decision being made and the right balance being struck.

Most design teams these days try to broaden the cost base of decisions. As Portman states, '… the *value* of any building is related to the long-term income from it, rather than to its basic construction costs'. However, capital costs will still be important to cost conscious clients and the capital costs are needed to get the project started and finished. Decisions must be influenced by the broader implications of the effect on the value of a project. This is difficult to judge *before* a project is complete. In the final analysis the cost planner will give the capital cost facts, life cycle cost forecasts and other cost data to the client and the designer, and they must judge the impact any change will have on the overall value of the project. It is not an easy decision.

There is no simple answer to the question of reconciling cost and value. It is a 'moving target' that changes with the type of development, location, time, client and market factors. We are making decisions that 'freeze' the variables for a moment in time (a snapshot) and we hope that events turn out the way we predict they will. The aim of this question and this chapter is to raise your awareness of these issues and to begin to develop a greater sensitivity to these questions when you are involved in what may appear to be simple and ostensibly straightforward decisions! The quote is also an important reminder that design team activities are more than just the interaction of narrow specialisations. Solutions are improved when all disciplines broaden their vision and appreciate the benefits that can be gained on behalf of the client and the wider community by considering factors beyond the limitations and constraints imposed by capital costs. And remember that a project consists of many constraints; cost is just one of them, but an important one.

Question 1.2: Statistical Publications

What are their main contents?
List the relevant publications for the construction industry in your area or region.

See the contents pages of each publication.

The data are presented in tabulated and sometimes in graphical format to aid understanding.

What do they tell us about the construction industry?
The statistics show us the facts and by comparison with previous periods indicate trends. Since there is a wealth of data contained in each issue, many newspapers, journals and some organisations summarise the published statistics into a more easily readable and digestible format. They indicate workload patterns, which may assist the design team in advising a client, when to build or when not to build!

Are they useful?
They are essential. These publications give the most detailed analysis of the construction industry in any country. Data are collected on a regular

basis and comparisons can be made that indicate movement in the various sectors. This is essential for government and private sector planners and decision-makers. Expansion of construction in a particular area or sector may encourage investment by developers, government departments or material suppliers. Facts are essential to guide managerial decision-making in any advanced economy and the construction industry is so important to the national economy that its characteristics must be traced on a regular and consistent basis.

Review Question

Weaknesses in the construction industry

- In the total process an inherent weakness is the separation of design development from the realities of construction practice. It contributes to lack of practical design and to separate groups whose aims rarely seem to merge. It divides the design team from the construction team. No wonder clients are seeking to use more integrated procurement methods! The large number of small building and sub-contracting firms makes it difficult to achieve a comprehensive and unified approach to improving performance and efficiency within the industry.

- Patterns of conflict seem to characterise relations in the industry. Contracts are based on adversarial relationships, not co-operation and the sharing of joint objectives. As a consequence the client's needs are often lost in inter-party and inter-disciplinary rivalry, which seems to contribute little to the outcome.

- In the construction field, contractors have a shoddy image based on illegal and collusive tendering practices and bribery; some corrupt activities are far too prevalent for the general public to hold the industry in high esteem.

- Inefficient and outdated practices tend to reduce the potential for improvements in productivity.

- A poor quality finished product.

- Delays caused by factors which are within the control of the design and construction teams and should not occur.

- Cost overruns, many of which become news items because they involve public money.

- Poor marketing and client relations.

- Inefficient organisation of subcontractors.

- Lack of attention to public concerns of noise, nuisance, danger and generally poor public relations.

Many of these have been noted in this chapter and specifically you should include some of the items listed earlier under the following headings:

- Features of the construction industry

- High cost of product.

- Time delays.

- Organisation and participants.

- Impact of technology.

- Poor risk assessment and management.

- Estimating practices which do not relate actual costs, productivity, market environment and other factors to bids being prepared.

- Problems.

Do not imagine that such practices are confined to one area; they are present in many other states and in many countries.

Can they be remedied?
They have to be (!) and some are at the present time. The clients in the private and the public sector, government, contractors' and subcontractors' representative bodies, professional bodies and Industry Agencies have been working hard to improve the situation over the past 10 years.

Some of the changes include:

- Multi-skilling of the trades, so that demarcation disputes between unions and trades are reduced.

- Integrated approaches to design and construction services for clients.

- Better education of construction managers and practitioners.

- Partnering between clients, contractors, subcontractors and suppliers.

- Pre-qualification criteria for consultants, contractors and sub-contractors adopted on many projects by central/federal and state/local governments and by a large number of private sector clients.

- Minimum standards for registration and business operations.

- Adoption of benchmarking for productivity and national and international standards in Quality Management and documentation.

- Better marketing of construction services, especially by the large contractors.

- Improved penetration of overseas markets.

These activities should not be mere window-dressing; they must have real substance and show progress through improved measurable performance. The situation is certainly improving, but there is a long way to go and more effort is required. However, great hindrance to progress is the disparate nature of the industry caused by the large number of small firms that was noted earlier in this chapter.

Chapter 2

The Development Process

Strategic advice to clients depends on a grasp of 'building economics'. That is, relating building costs to client's financial planning. To claim to advise clients on their interests without reference to costs is a fantasy – and a particularly easily exploded one. Design without costs is meaningless.

Wendy Prince (RIBA, 1992)

Chapter preview

Cost management does not take place in a vacuum. The activities of cost control and cost planning form part of a total process, called the development process, which conceives and delivers the final project to the client and to the market. Many external factors can influence the form and type of development that is permitted to be built on a particular site. In effect, the design solution is never unrestricted. It is carried out within a framework of constraints. Some constraints are onerous, others are less so. The problem for the design team is to balance the need for high-quality design and what all participants would like to do, within the constraints imposed by the economic, technical, topographic, social, political and administrative environments we all find ourselves in.

In fact, some of the most important factors influencing the form and type of development are not those created within the design team, but those often imposed by the external planning, community and building authorities. The planning and building requirements for a new development can influence a range of design decisions and outcomes that may have a direct effect on some of the following features of a building:

- type of building and works (performance);
- size, height and possible form of building (scope);
- materials, components and methods of construction (quality);
- total cost of project (capital cost and increasingly, life cycle cost);
- duration of project (time or programme).

The chapter will focus on the key stages or phases of a project and identify the key participants who play a role in the complex business of converting the germ of an idea from a concept into a working building project. This is known as the development process and it may take a period from less than a year to several years in the case of the large development projects.

DEVELOPMENT FRAMEWORK

The development process, in many cases, involves a great deal of time, effort and money. Any attempt to reduce the amount of resources to achieve the same (or even improved) result is to be encouraged. This is the aim of building economics, which should clearly align itself with the needs of the client and the wider community. Where conflict occurs between these two groups, it is still necessary to have the financial information and framework for making decisions.

The development process involves more than the execution of construction works. In fact, if the problem were merely one of constructing buildings, then more projects would be likely to succeed! Buildings are designed for a purpose, to house and accommodate people, activities and processes. Their fitness for purpose requires a major effort on the part of the design team, the client and the controlling bodies. The actual users of the facility may not be known, and the interpretation of their needs and behaviours can be the subject of extensive design and user studies.

A simple framework of the development process is broken down into four important time and activity stages:

- briefing (or feasibility)
- design
- construction
- handover and occupation.

The terminology for each stage can differ from professional body to related professional body and country to country. In fact, when one compares the UK to Australia it is interesting to see the slightly different terms for what are essentially the same stages.

Our focus in this chapter is on the economics and cost planning activities for building projects in the design stage of the development process. That is, using Table 2.1 as a guide cost planning will commence in the briefing (feasibility) stage and continue through to tender documentation. We shall not be directly concerned with the procurement stage in this text. Cost planning starts in a limited, but important form in the briefing stage, but the process begins in earnest as the project information expands through the subsequent stages. Whilst the process begins at the briefing stage at the inception of the project, it is unlikely to create enough project information for cost planning to be applied rigorously. Cost planning, or more correctly, cost control during subsequent stages of construction and occupation is also important, but the emphasis in this text is cost planning the building design.

Table 2.1 Design Stages in Two Countries: UK and Australia

Organisation	RIBA[1] (UK)	AIQS[2] (Australia)	RAIA[3] (Australia)
Stage	Pre-stage A Establishing the Need		
1. Briefing	Feasibility (a) Appraisal (b) Strategic Briefing	(a) Briefing: Inception Feasibility	Pre-design Briefing
2. Outline Proposals	(c) Outline Proposals	(b) Outline Proposal	Design Stage Conceptual Design
3. Sketch Design	(d) Detailed Proposals	(c) Sketch Design	Design Development
4. Tender Documentation	(e) Final Proposals (f) Production Information (g) Tender Documentation	(d) Tender Documentation	Documentation (Tender/contract documentation)
5. Procurement	(h) Tender Action		Tender

Sources:
1. Royal Institute of British Architects (2000) *The Architect's Plan of Work*, RIBA Publications, London, UK.
2. Australian Institute of Quantity Surveyors (2000) *Australian Cost Management Manual: Volume 1*, Australian Institute of Quantity Surveyors, Canberra, Australia.
3. Neighbour, K. (2000) *The Client's Guide to Construction Projects*, Royal Australian Institute of Architects, Melbourne, Australia.

The cost planning activities and techniques used during the design stages are quite distinct from those carried out during the construction stage. For that reason, our work ends with the acceptance of the tender at the procurement stage and hopefully, this confirms that our cost planning has succeeded.

Early Stages

Many authors, such as Flanagan and Norman (1983), Ferry et al. (1999), Ashworth (2002) and Jaggar et al. (2002), have indicated that the decisions made during the early stages of the design (briefing, feasibility or inception) carry more far reaching economic consequences than the relatively limited decisions which can be made later in the process. This is shown diagrammatically in Figure 2.1.

On a typical project through its life from briefing to occupation and use of the building, the most significant cost reductions are achieved during the early planning and design stages. Once the project progresses into the construction stage the opportunity to make significant reductions is greatly diminished.

Relationship between life cycle cost savings and timing of implementation

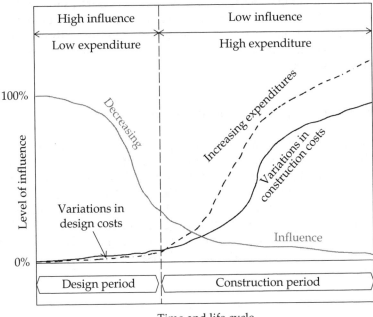

Figure 2.1 Ability to Influence Performance and Final Cost Over Project Life
Source: Adapted from US General Accounting Office (1978: 12).

Thus, the potential for reducing cost noted in Figure 2.1 is at its greatest
during the briefing and design stages at the start of the development stage
and lessens as the project moves into the succeeding construction and
occupation stages. There is also a cost investment needed to implement any
change. Any cost of implementation of a change must be taken into account
to calculate any *net saving* to the total project cost. As the project advances to
occupation by the owner/tenants the net saving potential of any change is at
its smallest. In fact, some changes may not be implemented because the net
effect is nil or negative cost to the project.

Whilst the emphasis in most design and cost studies is naturally on the cost
reduction possibilities, from our discussions in Chapter 1 we should also
appreciate the broader aspects of the project's aims and objectives. We should
recognise that improved performance in the trio of factors, time, cost and
quality, are equally, if not more, important to the client, users and the broader
community.

It is also worth noting that any alteration to the project will require changes
to the documentation (drawings/specifications/bills of quantities/schedules
and the like). These changes obviously increase in intensity and have greater

potential for disruption as the project proceeds. In the briefing and concept stages no major document revision is likely to be necessary as the project team is unlikely to have committed its decisions to detailed documentation. As more project information and documentation is produced through the life of the project then any changes will naturally have a greater effect on the documents already produced.

Pressures for and Purposes of Cost Management

The importance of cost control in building projects was thoroughly covered in the publication, *Cost Control in Building Design*, R & D Building Management Handbook, by the Ministry of Public Buildings and Works (UK) 1968. It was a programmed text, which is now out of print. However, Flanagan and Tate (1997) have published an updated version, which retains the original programmed learning approach to the cost planning material (see Further Reading for details). The following notes identifying the rationale for cost planning are largely summarised from, and based on the original text that retains a timeless relevance.

The control of building costs is now much more important than in the past because of four main pressures:

- *Pace of development in building is increasing*: With each generation the desire for greater efficiency, effectiveness and lower real costs increases. These increased demands have led to higher rates of urban development with time, cost and quality having a measurable value to a larger number of clients. These clients are less willing to tolerate any delays caused by redesigning buildings when tender costs are too high. Strategies and techniques are needed to produce designs for buildings to the required quality for a suitable cost and within an appropriate time.

- *Clients' requirements are becoming more complex*: The increasing technical complexity of modern buildings means that there are many more opportunities for the cost of a project to get out of hand. An effective system for the sensitive and accurate control of costs from inception to handover (and beyond) on a project is required.

- *Client organisations are larger*: Client bodies generally have tended to increase in size through mergers, expansion, takeovers of other companies and public ownership. Nowadays, very few large building programs are financed by one wealthy individual or company. The modern client can be a large industrial corporation, a statutory board, a major retailing chain, a government department or a national and multinational client extended over many states, countries and continents.

These large organisations are commonly directed by a board or committee that draws on specialist staff for the examination of expenditure. This has become necessary because of the complexity of the business and the volume of business that is undertaken. Often, there is a legal requirement that the board or committee has to satisfy the shareholders or the public that their money is being spent wisely. Thus, future expenditure must be forecast and controlled accurately.

The efficiency and accountability of these client organisations must be matched by the construction industry and the design professions, and especially by the project design team members. This is particularly important in the field of cost planning and control. Actual building expenditure must correspond closely to estimated expenditure. Cost planners must have the skills and experience to be able to forecast the cost of a project before detailed drawings and specifications are available, and must be able to keep actual costs within their forecasts. Client organisations expect and demand this level of professional ability.

- *Increased use of new techniques and materials*: When traditional estimating methods and systems of cost control were developed, the range of building types and forms was much more restricted than it is now. Many different building types are now required for a large number of functions, some are contained within the same project: airports, power stations, shops, hotels, apartments, laboratories, factories, warehouses and so on.

Up to about 1950, the range of building materials was much more limited. Most were local and only limited amounts were imported. A massive array of materials is now available to the design team including plastics, precast concrete, aluminium sheeting, fibreglass, plasterboard and composite materials. Similarly, new construction techniques are available to the contractor: curtain walling, pre-stressed concrete, tilt-up slabs, table formwork, industrialised/factory made and prefabricated units, and many other components and systems. The array of materials and techniques available in modern projects is dazzling, and often demands specialised skills from the design and construction teams.

With greater emphasis on sustainable design cost planners and designers will have to expand and revise their store of materials and construction knowledge to include more sustainable construction and design approaches. Some of these sustainable approaches may involve embracing some of the traditional materials, techniques, recycling and passive design, and low-energy systems.

The increasing range of choice of materials, construction techniques, types and forms of buildings, has placed more demands on traditional estimating to accommodate these techniques. The cost planner has to be aware that using new and innovative design approaches may make these techniques less reliable and subject to greater fluctuation between the initial estimate(s) and the final cost. Additionally, it has become more difficult to achieve a balanced and sustainable design that ensures value for money. In fact, too many so-called sustainable designs do not represent good value for money for the clients, the users or the community. It is still essential to subject the sustainable design to the same rigorous evaluation procedures that apply to more conventional design approaches.

Modern cost control systems are designed to focus on the proposed expenditure forecast for the project while at the same time attempting to also satisfy the first two aims of any cost planning system: creating value and achieving a balance of expenditure. These are represented in Figure 2.2.

The need for a cost planning and control system, which matches these demands and pressures in the modern interactive social, economic and technical environments, is overwhelming. Another important dimension not shown on this diagram is the pervading influence of technology, not only in its influence on construction techniques and management, but also in society generally and how it affects the economic and social systems. Whilst many of these pressures operate at a macro-level, few, if any, modern clients can afford to embark on a major project without cost being the prime criterion for evaluation. Thus, any design and construction team cannot ignore cost control.

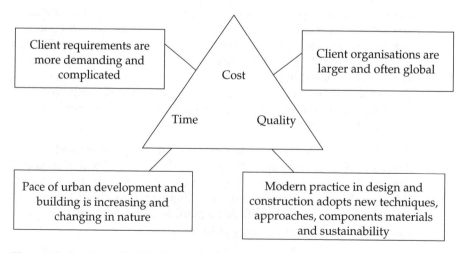

Figure 2.2 Pressures for Building Cost Control
Source: Adapted from Ministry of Public Buildings and Works (1968: 8).

The Development Process

It is important for you to be aware of the whole range of project activities from brief or feasibility (inception) to completion of the project, and on to occupation and the life cycle of the building. You can then appreciate the types of activities carried out during each stage and how they relate to other activities in the process.

The Royal Institution of British Architects (RIBA) *Outline Plan of Work 1998* (and also RIBA, 2000) defines the major work stages for the process of fully designing (and administering) a building project and provides the authoritative source for our framework for cost planning in this text. Similarly, the Australian Institute of Quantity Surveyors (AIQS) in Australia has defined cost control activities throughout the development process in its *Australian Cost Management Manual* (AIQS, 2000). As noted earlier in Table 2.1 the stages and terminology and activities at each stage are remarkably similar. Table 2.2 shows the RIBA Plan of Work in diagrammatic form.

Most countries have recommended development frameworks for planning, design, construction, handover and occupation of building projects. It is interesting to compare the RIBA (2000) version given in Table 2.2 with the AIQS (2000) model shown later in Table 2.3.

Table 2.2 RIBA Outline Plan of Work 1998

Pre-design			Design			
Inception or Feasibility			Pre-construction Period			
Pre-stage A	Work Stage A	Work Stage B	Work Stage C	Work Stage D	Work Stage E	Work Stage F
	1	2	3	4	5	6
Establishing the Need	Options Appraisal	Strategic Briefing	Outline Proposals	Detailed Proposals	Final Proposals	Production Information

Source: RIBA (2000).

Table 2.3 Australian (AIQS) Pre-construction Cost Management

Stage	(A) Brief	(B) Outline Proposals	(C) Sketch Design	(D) Documentation	(E) Tender
Output to Client	Brief Stage Budget	Outline Proposals Budget	Sketch Design Cost Plan	Tender Document Cost Plan	Tender Report

Source: Adapted from AIQS (2000: 1–2, 2–10).

Whilst they are similar in overall structure the terminology used for each stage slightly differs. The RIBA model is more comprehensive in its coverage, particularly of the early stages in *Pre-design Stage A* and in *Stage A, Options Appraisal* and *Stage B, Strategic Briefing*. The greater emphasis on these early stages seems to give recognition to the point made earlier in Figure 2.1 that the most critical project decisions are made during the formative stages of a project.

Table 2.2 identifies the various cost control stages of a typical project. The RIBA pre-construction stages from *establishing the need* at pre-stage A from prior to the brief to the tender action are clearly identified with their order and relationship to each other. Whilst the stages are shown in a linear relationship for simplicity, in practice some stages may have several iterations and possibly a backward step to ensure the correct decision has been made. For instance, intervention by one of the various planning and legislative authorities or by the client may cause a stage to be curtailed and commenced again with the new constraints identified. In an extreme case the design team may have to retrace its steps back to an earlier stage. So, whilst the above diagram shows a logical linear progression of activities and stages real life projects may not progress in such a smooth uninterrupted fashion.

Table 2.3 identifies the various cost control stages of a typical project operating in an Australian environment, from the quantity surveyor's perspective. The AIQS stages from the brief to the completion of the construction are clearly identified with their relationship to each other.

The important documentation in terms of output to the client is defined with the emphasis on the cost control documentation: brief stage cost, outline proposals cost, sketch design cost, tender document cost, tender report and final account. The flow of activities at each stage are summarised and the necessity to recycle or iterate the process and activities may also be necessary. Repetition of these activities to reach the correct and acceptable decision before proceeding to the next stage is essential for all the participants, especially for the client.

In fact, cost control or cost management extends into the construction stage (not shown in Table 2.2 or Table 2.3) and eventually into the occupation of the building coming under the heading of whole life cycle costing and facility management. As noted earlier, using the RIBA and the AIQS model frameworks we will be concentrating solely on the cost planning part of the pre-construction cost control stages.

With the integrated forms of procurement such as management contracting (in its various forms), relationship contracting, and design and build, cost control is still necessary as clients still need to know their commitment and the design team must work to a budget or cost framework. So, whilst some stages may be compressed in the models of cost planning noted by the RIBA and the AIQS cost planning should still begin at the early stages of briefing or feasibility stage, carry on into the design and documentation stages and continue into the construction stage in the form of cost control.

We have now introduced a model framework for cost planning based on UK practice described in the RIBA (2000) Plan of Work with recognition that the cost planning activities associated with this model also have application in other environments such as the Australian context (AIQS, 2000). In later chapters we will use this framework as the means for describing the cost planning activities, processes and outputs from each stage.

Participants in the Development Process

Creating a building design for a particular client in a specific location and implementing its construction ready for occupation is a complex process which involves many varied parties, bodies and organisations (The Aqua Group, 2003). The nature and scale of the project contributes to the complexity and number of parties who are involved in the process (Gray and Hughes, 2001).

How can we define and classify the various parties involved? Using a simple analysis we can allocate the participants into functional groupings or teams. However, there are always some individuals on certain projects who are unique and may defy classification!

The major groups or teams involved in a conventionally organised project may include the following members. The list is not exhaustive.

- *Client team*:
  - The 'Client' or 'Employer' or 'Owner' (an organisation, an entity or an individual)
  - Technical advisers (depends on project)
  - Financial advisers
  - Legal advisers
  - User group
  - Manufacturing/process advisers
  - Research and development group
  - Project manager
  - Planning supervisor
  - Property adviser (real estate)
  - Building/services maintenance division
  - Facilities managers

- *Design team*:
  - Principal designer, design leader, design team leader or architect (or lead consultant)

- Structural engineer
- Mechanical and electrical engineer
- Hydraulics engineer
- Interior designer
- Specialist designer(s)
- Quantity surveyor
- Other consultants
- Clerk of works/Site inspector

- *Construction team*:
  - Head, Principal or Main contractor
  - Site agent or foreman
  - Domestic subcontractors
  - Nominated subcontractors
  - Specialist subcontractors
  - Building suppliers
  - Specialist suppliers

- *Other groups*:
  - Local (planning and building) authority
  - Local community/neighbours
  - Unions
  - Politicians
  - Heritage and conservation groups and so on, depending on the location and nature of the project

The tendency over the years has been to create more specialisations as projects become more complex and beyond the capabilities of the established professionals in the client, design and construction teams. The role of the project manager has developed and this changing client culture is likely to mean that project managers will be a permanent feature of project arrangements (Fewings, 2005). In future years we can see this trend continuing and more specialists joining the teams noted here. For instance, with the present trend towards greater emphasis on client briefing and the early stages of the project, *project advisers, design managers* or *facility programmers* (using the American term) are likely to become more common. At the present time, *value managers* are becoming more prevalent on major projects, in particular, during Stage A, Options Appraisal, when many of the significant decisions are being made.

The list above is merely indicative of the type of advisers and personnel that may be found on a typical large-scale traditionally procured project. On individual projects the list may be expanded or contracted depending on circumstances. However, the role of the project manager to coordinate and integrate all the participants on behalf of clients in these complex projects is likely to continue (Latham, 1994).

Similarly, where alternative methods of procurement are adopted the design and construction teams may not enjoy a separate existence, as the important advantage of many of these arrangements is to integrate design and construction. Thus, the design and construction teams will become integrated as one team.

Under the traditional contract arrangement on medium to large projects as characterised by the teams above, these various groups have often pursued their interests without reference to the needs and desires of the other members or to the project itself. It is sad to say that too many building projects have been marked by conflict between the groups rather than co-operation. As a result, in the last decade clients and all the groups identified above have made strenuous efforts to overcome this lack of integration and co-operation by developing new project organisation arrangements, which closely focus on the 'real' project and client objectives, and not on a particular professional or other group's interests. The design team should aim to overcome some of these potential factional rivalries and to develop more efficient and effective project procurement methods (Ashworth, 2002).

The co-ordination and interaction of these various internal and external teams and groups requires a great deal of management skill. The development process involves all these participants, or multiple *stakeholders*, as they are now commonly called. The client with the person(s) responsible and connected with the decision to build is often involved in a great deal of activity and decision-making before the design team is appointed.

Each project may be different, but typically, the process begins with a stimulus or trigger (usually by senior management) that change is needed in the organisation's environment. It may be perceived that the organisation is operating ineffectively in its present facilities, there may be a need for increased efficiency, possibly a new market is opening up (expansion), introduction of new government regulations may force major change in present facilities or there may be a need to reduce the costs of production (Woodhead, 1999). Market research for the type of project being proposed when a detailed analysis of supply and demand for the functions or business proposed for the building(s) that are to be developed. It is essential that the design team and the cost planner are aware of these activities and can make a contribution if requested to do so. On many occasions the design team may not be involved in these activities and may not be called upon for advice until a site, or sites, are being considered and may have been acquired. Not a great deal of literature exists on these formative stages (conception, decision to build or project initiation) in the life of a project. One of the few authors to research this important area is Woodhead (1999, 2000) and Woodhead and Smith (2002).

The development process is involved and complex with a great deal of rework required and abortive work if the answer is negative or further work is needed. This is one of the riskiest stages in the whole life of a project. There are four important decision points in the process where the result will determine whether the project will continue in its present form:

- preliminary development appraisal (mainly economic);
- detailed viability study (preliminary economic, technical, functional);
- site acquisition or site suitability (preferably with designer's advice);
- application for planning permission.

Once the site has been purchased and planning permission has been achieved the risk profile of the project is reduced, but is not completely risk-free. For more information about the development process, see Darlow (1988), Fraser (1993), Millington (2000) and Ashworth (2002).

The programme of development then moves into the preparation of the design guidelines and Stage A, Options Appraisal of the Plan of Work and the appointment of the full professional team. This is where detailed briefing begins commences in earnest. As noted in Table 2.2 at the Pre-Design Stage A, a great deal of activity and decisions are needed before the design stages commence. The client has to establish client management team or management committee for the proposed project, appoint a client representative (project manager or architect) and appoint a cost consultant or cost advisor (usually a quantity surveyor). In addition, the client in conjunction with the emerging design team identify the objectives of the project, define the physical scope of project, establish the standard of quality of building(s) and services, forecast a realistic time frame for completion and occupation of the new facilities and most important establish a client agreed *budget*.

FACTORS INFLUENCING THE TYPE AND FORM OF DEVELOPMENT

There are a number of factors, some broad in nature, others more local, which can determine the purpose and function of the building and will influence the size, shape, scale and form of development. It is important to be aware that all developments operate within a framework of constraints involving activities, land use and land values, which in turn are influenced by a series of considerations that can establish whether a project can proceed and the form it will take.

The factors that can influence the form and type of development can be summarised as follows:

- Physical
- Legal/Administrative/Political
- Economic

- Technological

- Environmental

- Social.

In a totally free market system (that is, absolutely no control on development) for land and development it would be the unrestrained economic and social forces that would shape the physical form of our cities and urban areas. However, few, if any societies operate a completely free market system, and most are governed to a greater or lesser degree by the interactions between the various factors, most of which are constraints, noted above.

In some situations one particular kind of influence may be more important than in other situations. For example, physical factors may clearly predominate when deciding what land will be used for a marina or a steep hillside development, while economic considerations may be the most important when deciding what land is to be used for offices, hotels and shops. However, it should be noted that there are no situations where one factor acts totally alone.

Physical

This factor is taken in its broadest sense to include all those natural and man-made constraints to, and opportunities for, development, such as:

- Land

- Infrastructure

- Stock of Existing Buildings

- Capacity of Construction Industry.

Land characteristics
These are the geographical influences on land, including topography, climate, soil conditions, location, area, availability, cost and liability to earthquakes, flooding and cyclones. In the past, land characteristics have played an important role in new urban areas by influencing where cities have been established and the shape of their growth. Natural barriers to growth such as rivers, lakes, hills and mountains can still be a deciding factor in governing the overall shape of an urban area. However, modern technology (see later) can reduce the impact of these natural 'obstacles', but often at a price. In fact, for modern developments residential and other users often consider the proximity to natural features such as lakes, rivers and mountains, a desirable amenity and the developer has to consider the balance between the additional costs of these projects in difficult terrain and the extra benefit in terms of greater income in terms of sales or rental return.

Soil or geological conditions may have a bearing on the pattern of land use, the density and costs of development. Foundation type and complexity for any particular building are determined by the subsoil conditions. Unstable soil conditions can add costs to foundations and may discourage development, or some sites are not developed, as intensively as their location permits.

Climatic conditions can also play their part in influencing development. The direction of prevailing winds may play a part in the location of 'smokestack' and noxious industries. Generally, residential areas attempt to be located away from the detrimental effects of such industries and the planning authorities generally recognise this. Unfortunately, in many cities it is common to see low-income housing located close to the industrial area, whilst those who can afford to locate elsewhere attempt to avoid the wind-borne pollution. Micro-climatic differences can occur within relatively small distances and can determine whether an area is more desirable or undesirable depending on the environmental conditions created.

Topography can also influence building costs. Whilst elevation and views (particularly as development moves up a hill) are generally considered desirable, it comes at a premium. The cost of building on steep slopes is greater than that of building on level sites (Simpson and Purdy, 1984). It is likely that higher-income groups who can afford the additional building costs in these desirable locations are prepared to pay them in return for additional amenity and value of elevation and views. It is no accident that in most urban areas that more expensive housing is found moving up a hill or elevated location. The diminishing amount of land as we move up a hill also increases its scarcity or supply … but that introduces economic factors into the question!

Infrastructure
Under this heading are the basic utilities (water, electricity, gas, waste disposal and drainage) required to service the building and the all-important factor of transportation or accessibility.

In the past, the utilities to a site or location have received scant attention. It has been assumed by many developers that they have a right to receive these basic services irrespective of the location of the development. Increasingly the service providers, economists, urban planners, politicians and consumers who have to fund the expansion of the service system are questioning this. The unparalleled suburban growth (often called urban sprawl) of cities in the last 40 years have led many to question whether this urban expansion should be encouraged when many inner city areas are depopulating in people and jobs. As a simple and preliminary step to slowing this growth most service authorities now charge the actual cost (rather than a system average, or subsidised cost) for extending the utility network. That is, a 'user-pays' principle is now commonplace. This is a small, but necessary disincentive to the uncontrolled growth at the urban fringe. Existing urban areas commonly with diminishing population and under-utilised capacity in its services can offer lower connection fees, which in a small way can increase their competitiveness in terms of costs of location.

Issues of restraining urban sprawl and encouraging urban consolidation in inner city and existing areas have gained a great deal of attention over the past few years. It is a trend and a movement that is likely to gain momentum in the future and no developer can afford to disregard its ramifications.

The physical provision and quality of transportation facilities – roads, highways, freeways, bridges, tunnels, railway systems, tramways and light-transit systems, buses and even footpaths and bike-paths – plays an important part in increasing or decreasing the desirability of a particular location. In economic and urban planning terms transportation embraces the broader concept of *accessibility* and in an age of increasing personal technology accessibility in its physical forms may now be questioned. Access may not always mean close proximity with the use of the Internet and other communications and information technology.

Accessibility and infrastructure
Accessibility is considered to be more a more important aspect of location and home/work relationships, but it is obviously linked to the technology of its provision in terms of the physical facilities (infrastructure). It is not only concerned with the distance travelled between two places, but more importantly with the time taken to travel that distance. Accessibility is concerned with the real costs of transit and the real benefits gained from it. For example, several locations may create similar costs of transit, but a business is likely to be greatly influenced by the total number of consumers purchasing that firm's goods or services in a specific location (catchment).

Businesses and individuals value accessibility as a means of reducing time and costs and improving the benefits (social as well as economic) that can be gained from a specific location. Observations of any urban area indicate that with an existing transport system any movement will be concentrated along given transportation infrastructure. Therefore, sites and locations adjacent to the transportation links will have the advantage of greater accessibility, over those a greater distance away from the route. Sites at transport intersections (or nodes) will enjoy greater activity, and those at the centre of the transporta-tion system (usually the central activities area of the city) will command the highest levels of activity and greatest relative advantage over any other part(s) of the urban area. Thus, it is no accident that the accessibility advan-tage possessed by the city centre is a key factor in the patterns of urban land use and values.

In recent years this supremacy has been eroded by the appearance of suburban commercial centres, which have exploited the growth of modern cities, and they in turn have accrued the benefits of accessibility through their position on the transportation network.

In the future, concepts of accessibility, which up to now have been mainly centred on access to work, leisure and home, may have to be broadened and possibly redefined. New technology which now permits many activities (leisure and work) to be carried out successfully without having to physically transport yourself to a specific location may have a dramatic impact on the shape and form of our cities and upon the type of houses and workplaces we may build. Evidence of these changes is becoming apparent in many societies, although the 'old' physical principles of location still have greater potency.

Capacity of the construction industry
The characteristics of the construction industry were described in Chapter 1. Hillebrandt (1975: 27, 2000:191) provides the definition of capacity as:

> ... the maximum output which is attainable by the industry, within the limits of conditions considered acceptable at the time.

The capacity of the industry will change over time and will be influenced greatly by government policy, economic conditions and the social climate. Efficiency and productivity greatly affect the ability of the industry to respond to demand for construction. Governments in many countries have attempted to improve the capacity of the industry by improving its productivity performance through the establishment of agencies or boards with a brief to improve efficiency within the industry (Latham, 1994; Egan, 1998). An efficient construction industry allows urban development to take place with fewer resources (labour, materials, plant, finance and management) in a shorter time, and for a lower cost. Lowering the costs of development, whilst maintaining the quality of the product makes a positive contribution to economic and social welfare. In short, more schools, offices, hotels, shopping centres, dwellings and factories can be built at a lower cost in a shorter period of time. A flexible construction industry can respond to its environment much more rapidly. This has the effect of increasing its capacity.

Stock of existing buildings
The age, condition, efficiency (functional and operational), maintenance characteristics, location and accessibility of the stock of existing buildings will influence the level of urban development and redevelopment at any time. If existing buildings have positive characteristics (and an ability to be renovated and upgraded) then the demand for new development (new build) will be lower. When the existing stock does not satisfy the demands of modern society then the pressure for change and new development is likely to be higher. The pressure for redevelopment and demolition of existing buildings will increase.

Present day trends towards conservation, reuse, rehabilitation and preservation of existing buildings under the general heading of 'recycling' tends to reduce demand for new development and contributes to making better use of the community's resources and facilities. These 'existing' buildings must be capable of 'recycling' into a well-located, modern, efficient new use that matches the demand for that new facility. Extensive use of existing building stock in this way can reduce urban sprawl and the many negative aspects of new urban development on the fringe of the city.

Legal/Administrative/Political Environment

Legal/administrative

The institutional environment plays a key role in determining the location and form of development. It consists of the laws, regulations, organisations and institutions in our society that control and affect the use of land and its development and redevelopment. Many participants in the development process tend to view the legal and administrative system as a negative and constraining influence of their proposals. The regional and local institutions which control development have the power to grant the legal right to develop in the way the developers wish.

The powers that these bodies hold generally consist of:

- Compulsory acquisition
- Restrictions on land use
- Licensing/registration
- Health regulations
- Planning requirements
- Building regulations
- Occupational health and safety
- and many others.

Political

Central, regional and local governments may give direct or indirect incentives to different areas of the economy and to specific areas to improve their efficiency, productivity and performance. Such assistance may result in greater investment in the stock of plant and buildings, and have the effect of boosting demand for construction.

Economic

The economic framework in a market-orientated free enterprise system such as that found in the UK, Australia and many other countries is the most important and dominant in determining the strength and direction of urban development. Economic factors can, and often do, override other considerations. Competing market forces influence the supply of development land, and demand for property and construction.

The market price will be determined by the interaction and subjective valuations of buyers and sellers. In practice, any individual buyer or seller has to take the market price as given. As a result of this competition between buyers and sellers, the developer who makes the highest bid for a particular piece of land will secure the right to develop it. If the intensity of the existing use increases, or the land use changes to a new one, then the value of the land changes to reflect the new use.

As development sites become available this competitive process can be imagined as taking place on all of them. The economic framework largely explains the growth and change in cities, and why points of greatest accessibility and convenience in the city and suburban centres have the highest land values. It also accounts for why those uses, which cannot afford to bid the higher market price, get squeezed out.

Under the action of market forces land uses tend to become arranged in certain patterns that reflect the demands of consumers for the goods and services of the buildings/property/land as expressed in the market price for the land. Bids for land, in effect, order the pattern of urban development.

Technological

Improving technology in transportation and construction methods can affect land uses in several ways.

- Lower costs for accessibility may lead to more dispersed urban areas.
- Faster and cheaper buildings may result in lower building costs and an increased supply of units.
- Improved construction methods may encourage an increase in densities, particularly in commercial centres.

Improved technology has to be balanced against its cost, riskiness and timing. As a general rule, the newest technology is often the most expensive, and the longer and more extensively it is in use, the lower its costs become.

When reviewing and analysing cities and urban areas they are a reflection of the conditions in which they were built, their period of development and state of technology. As construction technology has improved over the years the construction industry can provide higher buildings, greater spans, deeper buildings with greater environmental control and a large range of facade treatments.

Environmental

Environmental factors and considerations are playing a greater role in influencing the nature and form of development, and in many cases, whether the development proceeds at all. Issues of *sustainability* are now being incorporated into planning schemes, community considerations, developer's plans and building designs. Due to the upgrading of environmental factors and sustainability issues in the community's priorities many authorities and developers are drawing attention to them in their new developments and marketing them as a positive feature. The public are now keenly aware of environmental issues.

The traditional environmental factors of pollution; noise, dirt, dust, smoke, chemicals and traffic congestion still remain the primary focus of the community and neighbourhood. The sustainability issues of waste reduction, resource efficiency and low levels of energy usage are assuming greater importance with time, and are likely to increase rather than decrease in the future. A modern developer would ignore these considerations at their peril!

A continuing theme of environmental considerations is contained under the *conservation and heritage* heading. Most societies are more aware of their history and natural areas, which are often embodied in existing buildings and undeveloped natural surroundings. As noted earlier, the new technology (higher buildings, greater densities) tends to supplant the old on existing commercial sites. Some of the greatest conflicts occur when new developments attempt to destroy or change the nature of an area. The new more dynamic profit-making uses tend to overwhelm and destroy the less profitable and more passive existing use often embodied in natural areas and historical buildings and precincts. Balancing the needs of the community and the desires of the developer requires a great deal of political and planning skills and may result in a less than satisfactory outcome to both parties. Sadly, compromise and realism are attributes rarely found in many of these intractable debates and conflicts.

Social

These are the practices, customs and traditions of our society, which affect the use of land. For example, the recognition of the ownership and use of private property, the importance we place on health and education, and increasingly in modern society how tastes and fashions change to create new or modified demands for goods and services.

Developments and individual designs must consider people (the users or consumers), their values and attitudes towards:

- Population trends (demographic patterns).
- Family (size, establishment, nature, fertility rates).

- Social values (privacy, private ownership, community attitudes, the young, the old, the infirm, the mentally handicapped).

- Culture (popular, traditional, national, local, elitist, sensitivity to local custom).

- Individual aspirations (self-fulfilment, individual v society rights and responsibilities, employment).

- Education (value, nature, attitude, traditional, life-long).

- Spiritual awareness (nature, types, pervasiveness).

- Taste and fashion (rapidity and nature of change).

Generally, these social factors combine with other factors such as the economic and administrative to influence the outcome. However, it must be noted that when these social, cultural or spiritual factors become dominant (as when mining or development proposals come up against respect for local and possibly indigenous customs and beliefs), then it is more often the case that the social factors can often override the other considerations (mainly economic) that may be used to justify the development.

Sensitivity to the non-profit values (social) of a community is the hallmark of a well-considered development proposal. In many cases the development proposal may benefit and become less controversial from a revised approach that takes into account the views of these important stakeholders.

Chapter Review

An understanding of the whole development process is essential if you are to operate as a productive design team member. You must appreciate the nature and type of activities that have taken place before your involvement, so that you can be aware of the forces and people that have shaped the project to date. Similarly, whilst we mainly carry out our cost planning during the design and documentation stages of a project and complete them at the tender stage, you must recognise that the decisions made within the framework of the cost plan and project documentation are carried through to the construction and occupation stages. A legacy of decision-making that may last more than 50 or more years!

It is also important to recognise that your work as a cost planner does not stand in isolation. It is part of a team effort involving many individuals; some of them are highly qualified professionals in their field of expertise. Joint effort and a common recognition that the overall project goals take a higher priority than the narrow interests of particular individuals must be followed conscientiously. Disputes and disagreements between members of the same team and other teams will dilute the success of the project. The resolution of conflicts and disputes between the various parties is essential to the successful completion of a project. While this may not appear to have a direct influence on the practice of building economics and cost planning, it is essential that anyone involved in the development process understands the various interactions between the parties and the effect they can have on the total cost and time of a project. Often quite minor disputes can expand to cause major disruption, delays and cost overruns. While the cost planner may not be able to influence events directly, it is essential for all parties (and especially the project manager) to be aware of the financial consequences of their actions, or inactivity.

Review Question

Compare the two frameworks of the development process given by the RIBA and the AIQS above and trace the common stages and activities in each. Are there any major differences?

References

Ashworth, A. (2002) *Pre-Contract Studies: Development Economics, Estimating and Tendering*, 2nd edition, Blackwell Publishing, Oxford, UK. Chapter 11: Procurement.

Australian Institute of Quantity Surveyors (AIQS) (2000) *Australian Cost Management Manual: Volume 1*, Australian Institute of Quantity Surveyors, Canberra.

Darlow, C. (1988) *Valuation and Development Appraisal*, 2nd edition, Estates Gazette, London, UK.

Egan, J. (1998) *Rethinking Construction*, Report from Construction Task Force, Department of the Environment, Transport and the Regions, London, UK.

Fewings, P. (2005) *Construction Project Management: An Integrated Approach*, Taylor and Francis, London, UK.

Flanagan, R. and Norman, G. with Furbur, D. (1983) *Life Cycle Costing for Construction*, Surveyors Publications on behalf of the Royal Institution of Chartered Surveyors.

Flanagan, R. and Tate, B. (1997) *Cost Control in Building Design*, Blackwell Science, Oxford, UK.

Fraser, W. D. (1993) *Principles of Property Investment and Pricing*, Macmillan, London, UK.

Gray, C. and Hughes, W. (2001) *Building Design Management*, Butterworth-Heinemann, Oxford, UK.

Hillebrandt, P. (1975) The capacity of the industry, in Turin, D. A. (editor) *Aspects of the Economics of Construction*, Godwin, London, UK.

Hillebrandt, P. M. (2000) *Economic Theory and the Construction Industry*, 3rd edition, Macmillan Press, Basingstoke, UK.

Latham, M. (1994) *Constructing the Team: Final Report of the Government/Industry Review of Procurement and Contractual Arrangements in the UK Construction Industry*, HMSO, London, UK.

Ministry of Public Buildings and Works (1968), *Cost Control During Building Design*, HMSO, London, UK.

Prince, W. (1992) Delivery added value, in *Strategic Study of the Profession – Phase 1: Strategic Overview*, RIBA Publications, London, UK.

Royal Institution of British Architects (1992) *Strategic Study of the Profession*, RIBA Publications, London, UK.

Royal Institution of British Architects (2000) *The Architect's Plan of Work*, RIBA Publications, London.

Simpson, B. J. and Purdy, M. T. (1984) *Housing on Sloping Sites: A Design Guide*, Construction Press, Longman, London, UK.

The Aqua Group (2003) *Pre-Contract Practice and Contract Administration*, Hackett, M. and Robinson, I. (editors), Blackwell Publishing, Oxford.

United States General Accounting Office (1978) *Computer-Aided Building Design*, US Department of Commerce, Washington, DC, USA.

Woodhead, R. M. (1999) *The Influence of Paradigms and Perspectives on the Decision to Build Undertaken by Large Experienced Clients of the UK Construction Industry*, Unpublished PhD Thesis, School of Civil Engineering, University of Leeds, UK.

Woodhead, R. M. (2000) Investigation of the early stages of project formulation, *Facilities*, **18** (13/14), pp. 524–534.

Woodhead, R. and Smith, J. (2002) The Decision to Build and the Organization, *Structural Survey*, **20** (5), pp. 189–198, Emerald Press, Bradford, UK.

Further Reading

Ashworth, A. (1999), *Cost Studies of Buildings*, 3rd edition, Longman Scientific and Technical, Harlow, UK.
   Chapter 3: Development appraisal, pp. 23–47.

Ferry, D. J., Brandon, P. S. and Ferry, J. D. (1999) *Cost Planning of Buildings*, BSP Professional Books, London, UK.
   Chapter 2: The starting point – developer's motivation and needs, pp. 8–15.
   Chapter 3: Framing the developer's budget, pp.16–32.

Flanagan, R. and Tate, B. (1997) *Cost Control in Building Design*, Blackwell Science, Oxford, UK.
   Chapter 1: The importance of control over expenditure, pp. 1–9.

Lean, W. and Goodall, B. (1966) *Aspects of Land Economics*.
   Chapter 9: Land use within an urban area, pp. 133–152.
   Chapter 11: Urban growth and change, pp. 173–192.

Morton, R. and Jaggar, D. (1995) *Design and the Economics of Building*, E & FN Spon, London, UK.
   Chapter 15: Commercial values and the property market, pp. 339–358.
   Chapter 16: Values, cost limits and prices: the case of housing, pp. 359–380.

Review Question: Development Process Frameworks

They are both very similar! It is not surprising that both frameworks are alike, since they both describe a similar cost planning process in whichever country they are located. The terms used may slightly differ, but even here there is a great similarity inception, feasibility and outline proposals are all the same and are used at the same time in the process. The term sketch design is replaced by detailed proposals in the RIBA Plan of Work, as does final proposals and production information supplant tender documentation in the AIQS model. The greatest difference between the two models is that the UK version has placed much greater emphasis and definition on the early stages of the project and the Briefing stage has expanded to include *Establishing the Need*, *Options Appraisal* and *Strategic Briefing*.

These idealised programmes of activities can be criticised for being linear and segmented in a way that practice is not, but they are useful as a framework for practitioners and other participants in understanding the logical development of a project. They are also used to define the fee stages on a project for the various participants!

Chapter 3

Design Economics

A man who knows the price of everything and the value of nothing
> Oscar Wilde
> Definition of a cynic ...

Chapter preview

The financial considerations of a project, which are partially controlled by cost planning procedures form part of a range of factors to be evaluated and balanced by the project manager on behalf of the client group. These primary project factors are defined in the brief and are categorised as:

- functional

- technical

- aesthetic

- financial and

- environmental.

Maintaining a balance between all of these factors requires a high level of project management skill. An imbalance, or undue emphasis in any one of the factors, is likely to result in an unsatisfactory project and low customer satisfaction.

These features are the characteristics of the project and are often considered and analysed under the simplified project factors of Time/Cost/Quality. They pervade the whole study and practice of cost planning. This relationship attempts to simplify the complexity of project factors or variables and it provides a useful analytical method for identifying the result of changes in any part of the design.

However, project cost decisions must be viewed in their widest context. At one extreme an undue emphasis on capital cost minimisation may result in a cheap and nasty building that neither the users nor the market considers desirable. Similarly, an undue bias towards the aesthetic appeal of a project may result in excessive costs, but with no commensurate increase in technical and functional utility or value. Many authors in project management have commented on this trade-off between:

- time

- cost

- quality (a broad category which includes the functional, technical and aesthetic aspects or distilled into scope and performance).

Time
The most obvious consideration for time is the time for construction. However, adopting a project management perspective time includes the time from the inception of the project to completion and an appreciation of the life cycle of the building through its occupational life. Similarly, the pre-construction (or pre-contract) stages of a project should also receive due consideration and this involves the stages in pre-construction some of which have already been mentioned; Establishing the Need, Options Appraisal, Strategic Briefing, Outline Proposals, Detailed Proposals, Final Proposals, Production Information, Tender Documentation and Tender Action. The pre-construction stage ends with the signing of the building contract by the successful tenderer (bidder). So this pre-construction time must be considered by the project manager in addition to the construction time to completion and total occupation of the building(s).

Cost
The capital cost seems to dominate this factor, but the design team more than ever these days has to consider a whole range of other costs in the total cost of a project. These other costs include:

- Acquisition costs (financial aspects of investment negotiations, professional fees, cost of land, legal charges, holding costs, authority fees and finance charges, and so on).

- Life cycle costs (maintenance, cleaning, servicing, renewals and repairs, rehabilitation including alterations and additions, annual charges for utilities or energy costs, fixed charges and other support costs).

- Operational costs (cost of salaries for employees, provision of amenities, machinery, power and cost of adapting a building to meet changed conditions).

- Demolition (although generally not a significant cost in most cases the disposal of whole buildings such as nuclear power plants or asbestos-ridden can cause serious environmental and cost difficulties and so this factor should not be underestimated).

Any design and cost planning focusing solely upon capital costs will have failed and the cost planner must make the design team aware of the life cycle cost consequences of their choices.

Quality
This factor is the most difficult to define as it encapsulates so many aspects of the completed project. For instance quality can refer to any or all of the following characteristics:

- function (fitness for purpose as a building and for each space);

- performance (weather exclusion, durability, strength and stability);

- aesthetics (appearance and appeal);

- technical (construction methods, materials and fire protection); and

- environmental (sustainability considerations).

The designers have the formidable task of integrating all these quality requirements into a project that combines them to provide the clients, users, customers and all stakeholders with a building satisfying the three conditions first expounded by Vitruvius in ancient Roman times, *commodity, firmness and delight* (Wotton, 1969). These fundamental principles still apply today.

The connection, proper integration and balance between each of the factors of time, cost and quality on any particular project provide it with its value. Sadly, this combination is not fixed and cannot be defined in advance or given in a formula. Rather it relies upon the skills of the design team to get the right mix.

These three factors are illustrated in Figure 3.1 below often termed the *objectives triangle*.

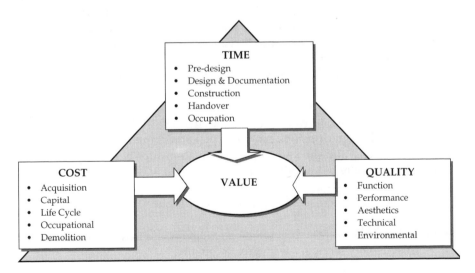

Figure 3.1 Time, Cost and Quality in a Project

The ideal balance between all factors is determined by the interaction of the various groups and the vigour and determination with which each defends and protects its interests. The role of the project manager in balancing and coordinating these interests is crucial. Many authors define project management in terms of controlling these three objectives in a project.

> ... the process of integrating everything that needs to be done ... as the project evolves through its life cycle (from concept definition through implementation to handover) in order to ensure that ... [the project's] objectives are achieved. Expected project outcomes or objectives which are common to most projects are the achievement of agreed time, cost and quality/scope/performance objectives ... on time, in budget and to specification.
>
> AIPM/CIDA (1995: 11, 12)

While much of the work of a cost planner concentrates on advising the design team on costs at the individual project level, it is necessary to be aware of the broader picture that embraces the other important factors of time and quality and how they may affect the total value of the finished project.

TIME/COST/QUALITY RELATIONSHIPS

Interaction and conflict consistently arise during the design and construction of a project between the time, cost and quality elements. In the past, and traditionally, the architect may have tended to favour better quality and higher cost. On the other hand, the builder when placed under commercial pressures has tended to favour lower cost and a timing of progress that optimises his resources. Sensitive project management and cost planning must allow these elements of time, cost and quality to be varied, interchanged and balanced in order to achieve the client's objectives. It is impossible to maximise or optimise all three criteria at once. To achieve a successful project a suitable balance must be achieved.

Detailed design decisions can have a major effect on all three factors: quality, cost and time. Design decisions cannot be divorced from construction without major cost and time penalties. The relationships between project factors can generally be simplified as follows:

- time usually increases with demands for higher quality;
- increases in time generally results in higher costs;
- Cost usually increases with greater quality.

The emerging pre-design cost planning, project management and construction management arrangements are developing the techniques and the information flows to allow suitable balances and 'trade-offs' to occur between these elements. The design team should be fully aware of the effect of its decisions on these three important factors.

Our discussion will concentrate on the cost elements of a project. The quality/performance factors are a complex and wide-ranging study lying outside the scope of this book. However, time, design or quality factors will be referred to where they clearly influence cost.

Cost of Construction

Speed of construction can be pursued regardless of cost and quality and with a total disregard of the final effects on the cost of a project. Cost control, in too many cases, can be pursued too single-mindedly at the expense of quality and time (and the relationships of individuals involved in the project!). However, if a client has a fixed upper limit or budget for capital costs, then cost control must dominate the three elements completely. The preferences of the client will obviously influence the priority given to each factor. The design team must assess the client's preferences in these areas to determine the priority given to each.

All parties must recognise that it is unrealistic to set the objective of getting the most building for the least cost. The objectives must be stated in a more reasonable way, such as obtaining the most building for a defined cost, or the lowest cost for a stated building area or downgrading cost by describing the problem as designing a building to satisfy a specified set of planning and functional criteria. However, it is an unfortunate characteristic of the design and building process that designers tend to give more emphasis to performance/quality objectives than to cost control. This makes the cost and time control processes inherently difficult for those engaged in it.

The other difficulty faced by the design team is that cost and time can be predicted with a reasonable amount of confidence. However, quality falls into an area where greater subjectivity and often strong opinions prevail. This makes it difficult for objective decision-making and this is where inspired and experienced project management is necessary.

Good cost control during construction is closely related to the care, thoroughness, and quality of the planning and decision-making process during the design stages. Changes to the design and the unexpected or unplanned generally cost more money. Therefore, it is better to concentrate the design team's efforts on making those decisions which greatly affect cost during the design stages and before construction commences. This makes cost control easier. Hurried planning and less well thought-out

Table 3.1 Proportion of Total Costs Spent on Building Services

Building Types	Outline Description	% Services of Total Building Cost
1. Administration, Civic	Offices, single storey	32.3
	Offices 2–3 storeys	31.8
	Offices 3–10 storeys	37.7
	City Hall, 2000 capacity	34.4
	Town Hall	32.5
	Library, suburban, single storey	31.8
	Library, suburban, 2–3 storey	32.7
	Art Gallery, city	34.4
	Law Courts	32.9
	Police station, suburban	34.3
2. Banks	Suburban, single storey	28.4
	City centre, 4–10 storeys	33.7
3. Educational	Primary school, single storey	17.5
	Secondary school, maximum 3 storeys	18.2
	University Lecture Theatre, 1–2 storeys	39.4
	University Science Laboratory	41.5
	University Animal Research Laboratory	44.9
	University Library	34.4
4. Hospitals	Hospital, general, multi-storey	52.1
	Multi-purpose health centre	40.3
5. Hotels	City hotel, 3-star, medium-high rise	35.0
	City hotel, international standard	36.7
	Serviced apartments, 3-star	35.1
6. Industrial	Warehouse, Low Bay, letting	15.9
	Warehouse, Low Bay, owner occupation	20.9
	Warehouse, multi-storey, maximum 6 storeys	36.5
	Factory, large span, owner occupation	20.6
	Factory, large span, offices, 2 storey	31.4
7. Offices	No air-conditioners, single storey, basic	19.1
	Air-conditioners, single storey, fully serviced	34.2
	4–7 storeys, fully serviced	37.4
	7–20 storeys, fully serviced	38.4
	21–35 storeys, fully serviced	33.7
8. Residential	Individual house, medium standard	15.2
	Individual house, prestige standard	22.5
	Flats, medium standard, maximum 3 storeys	21.6
	Flats, high standard, multi-storey	35.6
	Aged persons, 2–3 storeys	37.2
9. Retail	Suburban supermarket	35.0
	Regional Shopping Centre, medium standard	36.0
	City department store	42.7

Source: Analysis of Rawlinsons (2005: 63–107)

designs generally lead to more indiscriminate and aggressive cost planning. In this situation all parties tend to suffer.

Cost control also has to recognise the nature and generators of initial costs in buildings. This we now discuss as cost trends.

Cost Trends

A major cost trend in all countries has been that a greater proportion of building budgets are being spent on services within the building. This reflects the general rise in standards of living, the desire for more attractive and closely controlled internal environments, increase in technological and functional complexity and the need for greater energy conservation. For example, these cost trends are summarised in Table 3.1, which are taken from Australian data and represent the situation in most advanced economies.

Table 3.1 emphasises the point that building services are forming an increasing proportion of the total costs of a project, and secondly, that as standards of living increase clients, tenants, users and customers tend to demand ever higher standards of internal comfort and ease of access and travel through a building.

DESIGN DECISIONS WITH MAJOR COST CONSEQUENCES

This section cannot consider all the detailed design decisions that have to be made for specific projects and their effect on the final cost. Only those broad design or planning decisions, which have a major impact on the total costs of most commercial projects, are considered here.

Such decisions identified in the literature and practice are:

- total height
- storey height
- circulation areas
- plan shape (cost geometry)
- size
- cellular or open plan
- mechanisation/prefabrication
- redundant performance
- buildability (constructability).

These design decisions are now considered in turn.

Total Height

As a rough guide the fewer the number of storeys the more economical is the project (in unit cost per square metre). This is mainly because of the greater circulation and servicing requirements required for larger, multi-storey projects. This can also be seen in Table 3.1, where the greater the number of storeys, the larger the percentage spent on services in the building.

The high cost of sites in central city area means that developers must extract the maximum floor area from the premium land acquired. Thus, there is a tendency for central area sites to be the most expensive within an urban area and high-rise buildings tend to be concentrated around the most accessible central area. As buildings are built taller, vertical ducts for building services tend to increase in size.

Intermediate service zones are more likely to be required, and vertical circulation cores (lifts and stairs) form an increasingly higher percentage of the floor plan. In addition, very tall buildings on restricted sites often require special frame solutions, which add considerably to costs. Construction costs also increase and productivity levels fall because of the loss of working time caused when workmen move up and down the building. Similarly, materials and components take longer to place in their final position for fixing. More expensive site lifts, cranage and plant will be required to move people and materials. This will also increase the builder's project overhead costs (preliminaries).

Storey Height

Variations in storey height affect the cost of a building without changing the gross floor area. Those elements affected by changes in storey height are the only ones mainly having a vertical component:

- Staircase and lifts.
- Service runs and waste pipes – service intensive buildings such as hospitals, laboratories and prestige offices are more noticeably affected.
- Volume – the volume to be cooled and heated may require a higher capacity cooling and heating source.
- Ceiling void – small savings in height can accumulate to large differences in very tall buildings. For instance, if a modest 100 mm saving in the ceiling/storey height of a commercial high-rise building could be made on each floor, this would accumulate to a substantial saving in the total costs of the project. However, ease of maintenance and life cycle costs may be adversely affected by the restricted space.

This is a case of not solving a problem in one dimension, but understanding its effects on other aspects of the building and its use.

- Foundations – in high-rise projects the accumulative variation in dead load could affect the foundation design. Awareness of the thresholds for changing from one type of foundation to the next is essential knowledge for designers and the design team.

Circulation Areas

The high cost of building is causing design teams to give greater consideration to this aspect of the planning and design of a project. Circulation areas can account for between 10% and 50% of the gross floor areas of a building depending upon efficiency of layout and function.

Clients and cost consultants obviously prefer a minimum area of circulation space compatible with the satisfactory functioning of the building. The planning requirements provide a minimum for any given type of building. Skimping on circulation space can have a big effect on user satisfaction and perception of the building and this must also be borne in mind by the design team.

The proportion of space is often determined by factors other than the architect's designing ability, and these can be summarised as follows:

- *Standards of usage* – prestigious buildings such as banks, luxury hotels, offices and apartments tend to demand more generous circulation spaces.

- *Vertical circulation* – higher quality buildings usually have more lifts and wider or a larger number of staircases than the regulations require. Zoning of lifts in very tall buildings helps reduce the loss of space on the upper floors.

- *Horizontal circulation* – the type (mainly corridors) and standard of building will influence this to a great extent. With open planning circulation is merged with the usable space.

Plan Shape (Cost Geometry)

The shape of a building has an important effect on its costs. The more a building shape retreats from a square (the circle is the most efficient enclosure of space, but the cost of building circular work makes it too expensive) the more expensive the external envelope (i.e. external walls, windows and external doors) will become in relation to the remaining building cost. As a building becomes longer and narrower, or its outline more complicated and irregular, so the perimeter/floor area ratio will increase and be accompanied by a higher cost due to the increase in perimeter length.

The perimeter/floor area ratio (also known as the wall to floor ratio) measures the efficiency of enclosing space. The RICS (1969: 10) advises the wall to floor ratio is 'calculated by dividing the external wall area by the gross floor area to three decimal places'. The area of the external wall is measured on the outer face of the external walls and overall openings such as windows and doors. However, it excludes the area of gable walls, parapet walls and walls below lowest finished floor level. An extract from the RICS (1969) *Standard Form of Cost Analysis* with these and other definitions is included as an Appendix.

A smaller wall to floor ratio expresses the requirement for less wall area to enclose an equivalent area of floor space. Thus, a lower ratio expresses a more economic design from this design/cost aspect alone. A greater perimeter length is an important design variable because the perimeter affects so many other high-cost elements.

Thus, it can be seen that the design decision on the type of external wall and window or cladding system to be used is an extremely important one simply because the external envelope carries such a large proportion of the project costs and influences many other areas such as the framing system and service installations.

Although the commonly used curtain walling cladding systems found on high-rise buildings have problems of heat loss and solar gain, and often carry higher capital costs, they have an important advantage in being fixed to the outside of the structure. Thus, the space otherwise occupied by a solid external wall can be utilised as useable floor area[1] and lettable floor space and in this way gains some extra rental income or space for the owner or occupier. Again, this is an example of reviewing problems in a multi-faceted way to get the best decision and closer to increasing the value of a project.

The problems of environmental control arising from the use of curtain walling may be minimised by the use of special glass and double glazing and can further be reduced by careful orientation and use of sun-shading. The need to match ever-increasing energy standards also needs to be carefully considered.

An irregularly shaped building also increases costs for these reasons:

- scaffolding areas and associated costs are greater;
- setting out of the project is more complicated;

---

[1] Defined as the 'total area of all enclosed spaces fulfilling the main functional requirements of the building' (RICS, 1969:9). The floor areas are measured at floor level from the general internal structural face of the enclosing walls.

- Cleaning and maintenance (life cycle) is more extensive, has higher costs and is possibly more difficult.

Nevertheless, despite our rational assessment of building shape it has to be noted that in practice irregularly shaped buildings are more aesthetically pleasing and as a result may be more desirable to potential users and customers.

Question 3.1

Four building plans A, B, C and D are given.

Calculate the wall to floor ratios for each plan. Assume a wall height of 3 m in all cases.

Which one is the most efficient at enclosing space?

Are there any other factors which may affect the total cost of each plan shape?

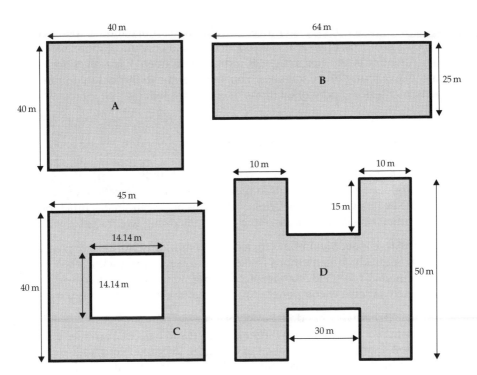

Figure 3.2 Four Building Plans (All Equal Floor Areas)

Size of Buildings

Increase in the size of buildings often produces reductions in unit costs. Reasons for this have been identified:

- Overhead or on-site costs are likely to account for a lower proportion of total costs. Certain fixed costs such as site accommodation, plant, compounds for storage of materials and components, temporary services, temporary roads may not vary appreciably with the size of the project.

- Benefits of the economies of scale in ordering materials and obtaining better prices from materials and components suppliers and subcontractors.

- If repetition of construction exists productivity gains may be made as a result of the beneficial effects of the *learning curve*. The learning and experience curve effect observes from practice and studies that the more often a task is performed, the less time will be required on each iteration (Ostwald and McLaren, 2004).

Cellular or Open Plan

Although in theory open plan offices should not be much more expensive than cellular offices, open plans in practice tend to encourage more expensive design/standards. Studies in the UK indicate that open plan offices are 6–8% dearer than cellular office schemes. This may be due to the fact that more expensive interiors and workstations may be adopted.

Mechanisation/Prefabrication/Standardisation

- *Mechanisation*: Increased mechanisation of building work can speed up production and can result in reduced costs of construction. On all sites the extensive use of prefabricated components eliminates much of the cutting on site and many of the time-consuming wet trades.

- *Prefabrication*: Maximum efficiency and economy can be achieved by mass production methods in factories aimed at producing large outputs of selected, standardised dimensional co-ordinated and interchangeable components suited to a range of building types.

- *Standardisation*: Standardisation is essential for component manufacturers to obtain larger series runs to balance turnover against capital, minimise time taken in changing over machines, reduce production costs by the bulk purchase of materials, and attain better quality production of fewer varieties of component types.

Standardisation of storey height permits the production of ranges of standardised units for staircases, refuse chutes, wall units and other components.

However, standardisation is a time-consuming, costly process. It is particularly expensive in research and development effort required to achieve solutions worthy of standardising and then to obtain agreement with the results. The prime factor in standardisation is repetition. Yet the normal way of working on building sites works against repetition. Just as a team works to its greatest efficiency, it is likely to be broken up and moved on to the next job.

Common standards of communication, components, construction and procedures in design offices, in factories and on sites could lead to the same high increases in productivity as have been secured in the manufacturing industry. Mechanisation, prefabrication and standardisation can only be adopted if designers are prepared to accept the design discipline the approach demands.

Redundant Performance

This is a useful term, which was coined by Bernard Williams, a UK Building Economist:

> Redundant performance in a building is money spent without benefit, commercial or architectural. The 'crime' can be committed at any level from the choice of materials to the shape of the building. The actual building cost, even when it is only a small proportion of the budget, is still far and always the most difficult part of the budget to control. Whereas the developer's skill and effort may save 1/2% on finance or 2% on fees, a skilful design team can really reduce the building budget. An unskilful team can, as suggested, make a large hole in the developer's profit.
> Williams's lecture notes at Brixton School of Building 1970.

It is helpful as a framework or guide to the design team and the client in considering whether to change or accept/reject any proposal. The approach is encapsulated in the designer label dilemma, which is illustrated in the problem of choosing a watch. Should it be a *Timex* or a *Rolex*? Both tell the time equally well in most situations, but there is a massive difference in cost! Why buy a car with a top speed of 150 mph (or 250 kph) when the speed limit is only 70 mph (110 kph)? In these cases the decision boils down to one of image and status, not one of substantially better performance warranted by the inevitable difference in cost of the two products. In the case of buildings it may be exactly the same motivation. Everyone should be clear that this is the case and there must be a clearly stated need, or a justification, to pay more for the material or products concerned.

Redundant performance is a useful guide to all designers, clients, cost planners and project managers throughout the design and construction process. The concept encourages a questioning approach to design decisions and continually prompts the client and the designer to consider alternatives that may still provide the same level of quality and performance for a much lower cost. That is, better value.

The redundant performance principle has been captured and formalised in the value management or value engineering techniques.

Buildability/Constructability

Buildability (or constructability, as it is more commonly known in the USA and Australia where the term extends into the engineering professions) has been defined by several authors and each contains slightly different emphases. However, the core principle is that it is an integrated approach where design takes account of the construction process and balances the various project and environmental constraints.

> Buildability is the extent to which the design of a building facilitates ease of construction, subject to the overall requirements for the completed building.
> Construction Industry Research and Information Association (1983: 6)

Whilst this may seem an obvious concept, for various historical and organisational reasons the construction industry has not followed this approach conscientiously. In the present day we cannot afford the luxury of ignoring the need to adopt buildability principles into our designs and activities. Thus, the design team has a continuing dual responsibility: designers should appreciate the process of construction when formulating their design; members of the design team with construction knowledge and experience should communicate this to the designers during the design development stages.

Buildability attempts to integrate design and construction into a better building that is easier to construct (Ferguson, 1989). Consequently, the project objectives of time, cost and quality should be more capable of being achieved. Good buildability contributes positively to all the factors in the following way:

- time – speedier construction times;
- cost – lower capital and life cycle costs;
- quality – improvements in the completed building performance and in maintenance characteristics.

In the area of costs, sound buildability will provide the client with financial benefits through the process of reducing costs and construction times to the builder. Research quoted by the Construction Industry Institute (CII) Australia (1992) indicates that where contractors have implemented a constructability program savings of an average of 6.4% have been made. More important is that this research also indicates that investment in management systems, which emphasise constructability show cost benefits as high as 50:1 for the extra investment made.

Identifying these lower costs and quantifying them for cost planning purposes is difficult during the design development stages for two main reasons:

- Whilst reduced costs can be shown in previous projects adopting good buildability principles, they cannot be guaranteed for the project in hand.

- Adverse client and design team decisions during design development and construction may negate progress towards greater buildability and lower costs. These decisions may be created by factors beyond the control of the participants.

In short, during the design development there are risks that have to be recognised and wherever possible allowance made for their effect and influence on the final outcome. Nonetheless, where buildability is being pursued thoroughly by the design team then it is reasonable that the expected efficiencies will be reflected in the final costs of the project. See Gray and Hughes (2001: 128–130) buildability checklist to guide design teams particularly during the early stages of the project. There may be situations where costs of previous projects using buildability are being used for cost planning purposes and in these cases they may be repeated. However, where costs for a conventional project are being used as the cost source, then the adoption of buildability should result in a *discount* being applied to the rates. This discount will depend on circumstances, but experience indicates that it may save up to 7–8% of the conventional costs.

Traditional methods of procurement of projects have not encouraged the integration of design and construction activities. Design and construction are divided by the contract arrangements and activities of the disciplines normally involved. Therefore, it is not surprising that procurement methods (such as *design/build, management contracting* and *construction management*) which link design and construction more closely appear to be gaining in popularity, particularly in Australia and the UK. There are other factors which account for some of the increased popularity of these procurement methods, but the message is clear to the design team: clients expect the integration of design and construction; it not an optional extra! However, there is no doubt that the measurable benefits gained by the time, cost and quality factors through the increased levels of buildability contributes greatly to the rise in popularity of the alternative procurement methods.

In situations where more integrated procurement methods are expected to be used, the division between the various teams is not so clear and the selected contractor will be part of the client/design team advising on costs. Nevertheless, the client should retain an independent cost adviser to ensure that any reported costs are assessed before they are accepted. In these situations the cost adviser/planner can measure any costs presented against conventional and *discounted* costs from similar projects. If the integrated project costs (and time) being presented do not indicate the level of savings expected, then questions should be asked of the selected contractor.

Steps for achieving good buildability have been recommended by CIRIA (1983) and they provide a sound framework for the design and planning activities for the design team on any project.

- carry out thorough investigation and design;
- plan for essential site production requirements;
- plan for practical sequence of building operations and early enclosure;
- plan for simplicity of assembly and logical trade sequence;
- detail for maximum repetition and standardisation;
- detail for achievable tolerances;
- specify robust and suitable materials.

That is, they represent good practice and common sense.

Chapter Review

Chapter 2 considered the development process. This chapter moves into closer detail and focuses upon the design decisions that affect the economics of the project. It is essential that the cost planner appreciates that the cost planning function does not concentrate solely upon the capital costs, but is also carried out within a project management framework that also embraces time and quality. The cost factor must also not be too narrow and should be broadened to include life cycle cost implications of decisions together with time and quality and the pursuit of higher project value. Within this framework the role of the cost planner is much more demanding and requires a great deal more involvement in the development and design process. It also requires a sensitivity and understanding of the full consequences of a decision.

Design decisions with major cost consequences were introduced in this chapter and these theoretical approaches must be further developed and refined with practice and experience in a project environment. The decisions presented and discussed here are not laws that are fixed or absolute, but must be applied with consideration of specific project factors.

In a similar vein it is necessary to make the design team aware that design decisions do not end with the preparation of sketches, drawings and other pre-tender documentation in the design stage. They have to be implemented on site by the contractor and subcontractor. These decisions also have to be supervised by the design team and they have ramifications on the cost, time and quality during the construction stage. The aim of this chapter is to develop and encourage a greater awareness in the cost planner of the consequences of making a particular decision and following a specific course of action. The factors that influence the implementation of a design decision should be continually surveyed, reviewed and appreciated by the design team.

Review question
Design and management decisions with major cost effects

Imagine that you are the cost planner in a design team with a commission to prepare preliminary costs for a new multi-storey residential apartment block in the Central Activities District of a major city. In your early discussions with the architect and other consultants, which of the design variables (some identified in this chapter) should be considered by the design team as having a major influence on the cost of this particular type of project?

Which parts of the design need particular attention?

Suggest, in broad terms, the type of effect they may have on the design solution(s).

References

Australian Institute of Project Management and Construction Industry Development Board (1995) *Construction Industry Project Management Guide for Project Sponsors/Clients/Owners, Project Managers, Designers and Constructors*, AIPM, Canberra, Australia.

Construction Industry Institute (CII) Australia (1992) *The Constructability Principles File*, CII Australia, Adelaide, Australia.

Construction Industry Research and Information Association (CIRIA) (1983) *Buildability: an Assessment*, Special Publication 26, CIRIA, London, UK.

Cordell Building Information Services (quarterly) *Commerical/Industrial Building Cost Guide*, Reed Publishing, St Leonards, New South Wales, Australia.

Cordell Building Information Services (quarterly) *Housing Building Cost Guide*, Reed Publishing, St Leonards, New South Wales, Australia.

Ferguson, I. (1989) *Buildability in Practice*, Mitchell Publishing, London, UK.

Gray, C. and Hughes, W. (2001) *Building Design Management*, Butterworth-Heinemann, Oxford, UK.

Laxtons Building Price Book (2006) Johnson, V. B. (editor), Elsevier Press, Oxford.

Ostwald, P. F. and McLaren, T. S. (2004) *Cost Analysis and Estimating for Engineering and Management*, Pearson Education, Upper Saddle River, NJ, USA.

Rawlinsons (2005) *Rawlinsons Australian Construction Handbook*, Annual Publication, The Rawlinson Group, Sydney, Australia.

The Royal Institution of Chartered Surveyors (RICS) (1969) *Building Cost Information Service (BCIS) Standard Form of Cost Analysis: Principles, Instructions and Definitions*, RICS, London, UK.

Spon's Architect's and Builders' Price Book (2006) Davis Langdon (editor), E & F N Spon, London.

Wotton, H. (1969) *The Elements of Architecture*, Gregg, Farnborough, UK.

Further Reading

The following books also contain material that is essential reading for the enthusiastic cost planner. Whilst they all cover similar ground each author gives a slightly different emphasis that will assist newcomers and practitioners in the field to understand the principles and concepts espoused.

Ashworth, A. (1999) *Cost Studies of Buildings*, 3rd edition, Longman Scientific and Technical, Harlow, UK.
     Chapter 6: Design economics, pp. 102–124.

Bathurst, P. and Butler, D. A. (1980) *Building Cost Control Techniques and Economics*, 2nd edition, William Heinemann, London, UK.
     Chapter 10: Economics of planning, pp. 57–67.

Ferry, D. J., Brandon, P. S. and Ferry, J. D. (1999) *Cost Planning of Buildings*, 7th edition, Blackwell Science, Oxford, UK.
     Chapter 10: The client's budget, pp. 91–108;
     Chapter 15: Cost planning the brief, pp. 197–212.
     Chapter 16: Cost planning at scheme development stage, pp. 213–236.

Morton, R. and Jaggar, D. (1995) *Design and the Economics of Building*, E & F N Spon, London, UK.
     Chapter 9: Concept, shape, plan: morphology and cost, pp. 185–220.

Stone, P. A. (1983) *Building Economy*, 3rd edition, Pergamon Press, Oxford, UK.

SUGGESTED ANSWERS

Question 3.1: Four Building Plans

Wall to floor ratio calculations
The assumed storey height of each plan is 3.0 m. In measuring the envelope area no distinction is made between the areas of windows, external wall and external doors.

The calculation of the various areas and perimeters leading up to the computation of the wall to floor ratio is shown below.

Comparisons	Plans			
	A m^2	B m^2	C m^2	D m^2
• Gross Floor Area	1600	1600	1600	1600
• Perimeter	160 m	178 m	230 m	260 m
• Area of Envelope ($\times 3.0$ m Height)	480	534	690	780
• Wall to Floor Ratio	$\frac{480}{1600} = 0.30$	$\frac{534}{1600} = 0.33$	$\frac{690}{1600} = 0.43$	$\frac{780}{1600} = 0.49$
• Cost of Walls (Assumed @ £500/m^2)	£240,000	£267,000	£345,000	£390,000
• Cost Index	100	111	144	163

Most efficient plan
The comparison above shows the great difference in areas and costs of external envelope that can arise on alternative plans with the same floor area. However, the envelope of the building is not the only cost consideration. Assuming a single-storey building the substructure, frame and roof costs can also have a large bearing on the total costs. Nevertheless, in these examples, even with a broader cost evaluation, Plan A would still retain the economic edge over the other plans. Its perimeter length, which has the greatest effect on all the elements noted above, is still the smallest.

Other factors
However, the building services (lighting, heating, air-conditioning instal-lations in particular) would probably show a different trend to that noted above. Plans B, C and D have relatively easy access to natural light and

ventilation afforded by the larger perimeter lengths in each plan. In contrast, internal spaces of Plan A have a more limited or non-existent access and may only gain a satisfactory internal environment by installing mechanical ventilation or air-conditioning. This may result in a significant increase in the total costs over the other layouts. It must also be recognised that in extremes of hot and cold weather the solar heat gain and heat loss of the longer perimeter may be a disadvantage.

Internal divisions of space can also have a great effect on the total costs. Extensive subdivision of space will affect the cost of internal walls, partitions, wall finishes and services.

This review is only made in broad terms. The type, form and location of the building are other important variables to be considered in any comprehensive cost review. If the plans were for multi-storey buildings, this would bring an important new dimension to the review process, requiring transportation, circulation areas and percentages, all needing close attention.

These aspects are discussed further in the review question.

Review Question: Design and Management Decisions with Major Cost Effects

Design variables to be considered
This chapter identified and discussed some of the major design variables to be considered by the design team on our apartment block project that will need particular attention. We will review:

- size

- shape

- height and total height

- circulation space:
  – lift lobbies
  – corridors
  – service spaces
  – number of lifts

- car parking:
  – surface (cheap)
  – underground (expensive in cost and time)

- external works

Cost significant elements and their effect
In the absence of any office data on this type, or similar types of project, refer to standard published pricing books (UK: *Spons*, *Laxtons*, Building Cost Information Services and in Australia: Ra*wlinsons*, *Cordell's*) for elemental breakdowns of high-density residential multi-unit projects. The high-cost elements can easily be identified. The following elements should be considered:

- *Substructure.* These are likely to be complex foundations such as piles; possibly requiring underpinning if adjacent properties are involved. Are basements necessary?

  On multi-storey projects such as this, the foundations costs are spread over a relatively larger floor area and often only represent a small percentage of the total cost; in the region of 1–5% of the total. The foundation type, subsoil conditions, depth and the need for basements influence costs. If basements are needed then the cost of the substructure will increase greatly.

- *Frame (columns, upper floors).* These are relatively high-cost elements representing between 10% and 20% of the total costs. The choice of frame type may not greatly influence total costs, but is likely to influence time. The frame cost is also affected by the shape of the building (See Question 3.1).

  Stairs may be considered as part of the frame, but they are not normally a cost significant item. Their cost is closely related to fire and evacuation requirements and therefore cannot be reduced below a certain minimum standard. Cost planning may influence the specification, but not the quantity to any great extent. On most multi-storey projects the cost of stairs would only be a minimal 1–2% of the total cost.

- *Envelope (external walls and windows).* As noted in Question 3.1, the shape of the building influences the cost of these elements. The specification (quality) is also important. The envelope can represent between 10% and 20% of the costs. It is therefore an important factor to consider.

- *Roof.* On a multi-storey project roof costs are not usually notable. The roof has less significance as the number of storeys increases. For instance, the roof on a 10-storey building of a similar floor plan is the same as for a 20-storey building, but the cost is distributed over double the gross floor area.

- *External Doors.* This item is not usually cost significant, although it may be the subject of a great deal of design activity.

- *Internal Subdivision (Internal Walls, Screens and Doors).* These can represent between 5% and 10% of the total costs. In an open plan office building with tenants taking responsibility for partitions; this is not likely to be a significant cost item. However, in residential blocks of this kind the design and layout of each unit is of critical importance. Significant savings can be accumulated through the number of units for what may appear to be modest layout changes. Open plan layouts with minimal internal walls/screens can grow into important cost savings when multiplied by a large number of units.

- *Finishes (Wall, Ceiling and Floor).* Wall finishes are related to the design decisions in internal subdivisions. Ceiling and floor finishes are more closely related to the quality selection. The choice of suspended ceilings may be expensive, but their selection may improve services provision and accessibility.

- *Fittings.* These are an important consideration representing between 5% and 10% of the total costs. The amount depends on the extent and quality of the provision. In residential projects, wardrobes, bathroom, kitchen and built-in fittings when multiplied by the number of units can be a significant cost item.

- *Services.* Consistently services rate between 30% and 40% of total costs and they are therefore a very important aspect to consider. The type of provision must be carefully evaluated at the design stage. Many heating systems are low in capital costs, but may prove expensive in running costs. However, a developer who sells the units upon completion may be less concerned with this aspect.

  The other major question is whether the developer or the owners provide air conditioning. Air conditioning provision in multi-storey residential may be expensive, but the developer has to consider the selling value of such a feature. Tenant responsibility for air conditioning may result in a multiplicity of different types of unit air conditioners, which can present a disunited and ugly appearance.

- *External Works.* These are not usually a cost significant item; probably providing less than 5% of the total costs. However, on a site with a large area and extensive treatment the costs could be substantial.

- *Preliminaries.* These are an important cost item, often representing between 8% and 15% of the total. Preliminaries are influenced greatly by factors such as access to the site, storage, height of the building, scaffolding, adjoining properties, insurance, supervision, temporary works, plant and equipment requirements and the complexity of the project.

The design team must consider all these aspects carefully. It is the responsibility of the cost planner to make everyone aware of the implications of the decisions that they make, in terms of the capital costs, life cycle costs and possibly the value consequences of any decision.

Chapter 4

An Overview of Cost Planning Stages

Cost planning is a system of procedures and techniques used by quantity surveyors. Its purpose is to ensure that clients are provided with value for money on their projects; that clients and designers are aware of the cost consequences of their proposals; that if they so choose, clients may establish budgets for their projects; and that designers are given advice which enables them to arrive at practical and balanced designs within budget.

The Property Services Agency, Department of the Environment (1981)

Chapter Preview

Cost planning, as part of a cost management framework is a total system that requires commitment from inception to the completion of a project. Cost planning implies a framework of procedures and demands a commitment from the design team to work closely with the cost planner to achieve the project's cost and other time and quality objectives. Cost management is a project and design team responsibility focused on delivering to the client the best value for money.

This chapter lays the theoretical foundations for the activities involved in the cost planning process. An excellent description of cost planning is given by Seeley (1996: 13) that summarises the aims of cost planning admirably:

> Cost planning aims at ascertaining costs before many of the decisions are made relating to the design of a building. It provides a statement of the main issues, identifies the various courses of action, determines the cost implications of each course and provides a comprehensive economic picture of the whole. The architect and the quantity surveyor should be continually questioning whether a specific item of cost is really necessary, whether it is giving value for money or whether there is not a better way of performing a particular function.

The theory of cost planning forms part of the wider concept of the cost control of a building project.

In this text, we concentrate on those cost planning activities, practices and techniques undertaken during the pre-tender design stages. We will be using the RIBA Plan of Work (2000) as the basis for our cost planning analysis and descriptions. There are six pre-tender work stages (A to F) and one Pre-stage A, *Establishing the Need*, which is an important stage and one that we will consider in our review of the cost planning process. The work stages defined by the RIBA are shown in Table 4.1 and a brief summary of the activities of each stage is also given.

Table 4.1 RIBA Pre-tender Work Stages in Outline Plan of Work

	Pre-design		Design			
	Inception or Feasibility				Pre-construction period	
Pre-stage A	Work Stage A	Work Stage B	Work Stage C	Work Stage D	Work Stage E	Work Stage F
	1	2	3	4	5	6
Establishing the Need	**Options Appraisal**	**Strategic Briefing**	**Outline Proposals**	**Detailed Proposals**	**Final Proposals**	**Production Information**
			Activities in each Stage			
Client has to establish client management team, appoint client representative, appoint cost consultant. Identify objectives, physical scope of project, standard of quality of building(s) and services, time frame and *establish budget.*	Identify client's requirements. Possible constraints on development. Prepare cost and VM studies to enable client to decide whether to proceed. *Cost of preferred solution.* Select the probable procurement method. Instruct development of preferred solution to strategic brief stage. Prepare outline business case.	Prepare strategic brief. Confirm key requirements and constraints. Identify procedures, organisational structure and range of consultants and others to be engaged for the project. *Target cost and cash flows.* Whole life costing.	Evaluate strategic brief with a consideration of time, cost, risks and environmental issues. Establish design management procedures. *Prepare initial cost plan,* project programme, cash flow. Develop project brief. Estimates of cost prepared.	Evaluate outline proposals, complete and agree user studies. Complete and sign off on project brief. Receive design and cost input from team and develop detailed design solution. *Firm cost plan and cash flow projection.* Development control submission. Review procurement advice.	Sanction and complete final layouts. Co-ordinate all components and elements of the project. *Cost checks design against cost plan.* Decide on procurement method. *Review design and cost plan.* Prepare submission for statutory approvals. Design is now frozen.	Prepare all co-ordinated production information including location, assembly and component drawings, schedules and specifications. Applications for statutory approvals completed. Provide all information *for final cost checks of design against cost plan.*

Source: RIBA (2000)

References to capital cost and cost planning in the activities shown in Table 4.1 are italicised for emphasis. It is noteworthy that all stages have a cost dimension and no work proceeds without the cost aspects being considered. Whilst costs receive recognition in the budget in *Pre-stage A, Establishing the Need*, in Stage A, Options Appraisal and Stage B, Strategic Briefing, cost planning begins in earnest during Stage C, Outline Proposals stage, where the initial cost plan is produced.

Note the italicised words in the activities in each stage in the lower section of the table. These words have been emphasised because they capture the essence of the cost planning activities in each of the stages of the RIBA Plan of Work as follows:

- Pre-stage A: Establishing the Need – *establish the budget.*
- Stage A: Options Appraisal – *cost of preferred solution.*
- Stage B: Strategic Briefing – *target cost.*
- Stage C: Outline Proposals – *prepare initial cost plan.*
- Stage D: Detailed Proposals – *firm cost plan.*
- Stage E: Final Proposals – *cost checks, design against cost plan.*
- Stage F: Production Information – *final cost checks of design against cost plan.*

In Part B we shall describe these cost planning activities in more detail.

The AIQS (2000) has also diagrammatically outlined the pre-construction cost control activities in the design stages of a project in its *Australian Cost Management Manual* (2000). Figure 4.1 builds on earlier diagram shown in Table 2.3. The four reporting stages of cost planning are clearly identified:

- Stage A – brief
- Stage B – outline proposals
- Stage C – sketch design
- Stage D – documentation.

These stages will be discussed in relation to the RIBA Plan of Work. It is important to note that the AIQS model in Figure 4.1 has a cost planning focus, whilst the RIBA model has a more comprehensive view of the role and responsibilities of everyone in the design team. Naturally, in this text we will be giving emphasis to the cost planning functions and activities within the context of the RIBA model.

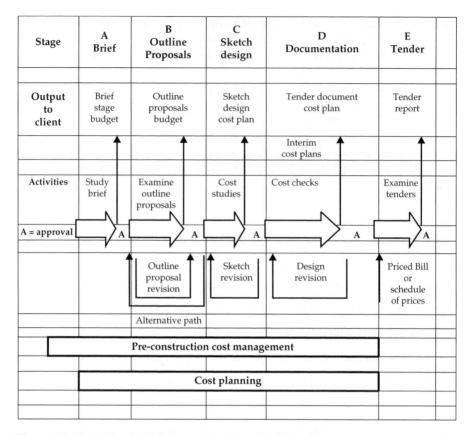

Stage	A Brief	B Outline Proposals	C Sketch design	D Documentation	E Tender	
Output to client	Brief stage budget	Outline proposals budget	Sketch design cost plan	Tender document cost plan	Tender report	
				Interim cost plans		
Activities	Study brief	Examine outline proposals	Cost studies	Cost checks	Examine tenders	
A = approval	A	A	A	A	A	
		Outline proposal revision	Sketch revision	Design revision	Priced Bill or schedule of prices	
		Alternative path				
	Pre-construction cost management					
	Cost planning					

Figure 4.1 Australian (AIQS) Pre-construction Cost Management
Source: Adapted from AIQS (2000: 1–2 and 2–10)

The AIQS cost planning framework in Figure 4.1 is also output focused and each stage forms the basis of reporting the latest costs to the client at a distinct and important stage in the design process. The output to the client will consist of a cost report, which must be approved before the project can proceed into the next stage. If the costs are not approved then the design must be reconsidered and possibly revised back to the start of that stage until the client is satisfied with the outcome. If the client is not satisfied with the outcome the project may be terminated.

The alternative path shown on Figure 4.1 from the end of Stage A directly to the start of Stage C (effectively missing Stage B) should only occur with the client's approval and when the design has been rapidly advanced from the briefing stage into sketch designs without the need to consider a large number of alternatives. This may occur on building programs with a consistent and a relatively standard approach to the design solution, such as schools, police stations or simple medical facilities.

Before we embark on a discussion of the theoretical underpinning to the cost control process during the design stages, it is necessary to be aware of the terms and vocabulary of cost planning by defining some of the important terms used in cost planning.

These are given at the start of this chapter because when any term or phrase is used which is not immediately defined, you can turn to this list for assistance.

GLOSSARY OF TERMS COMMONLY USED IN COST PLANNING

The following terms are crucial to your understanding of the theory and practice of cost control during the design stages of a project. They are terms used by cost planners and quantity surveyors and it is important that all members of the design team understand their meanings. It is important to understand the *new* vocabulary of cost planning, but remember its purpose is not to create a professional jargon that aims to exclude the uninitiated! Its aim is to provide a common and unambiguous language for design and construction team communication.

Budget   One of the most neglected terms in cost planning, but one of the most important. The AQUA Group (2003: 46) summarise the three primary ways that this critical figure is determined:

1. Prescribed cost limit – limit of capital expenditure is set down in the employer's brief, the amount normally being based on the maximum amount that can be afforded.
2. Unit cost limit – the limit of capital expenditure is calculated by the application of standard costs on a unit basis.
3. Determined cost limit – the limit of capital expenditure is based upon the economic viability of the proposed development.

Cost   Means price to the client in this context. It is *not* the internal costs of the Contractor. See Chapter 1 for a discussion of cost and price. We will retain the term *cost* in cost planning because of its traditional usage.

Cost Planning   A method of controlling the cost (price to client) of a project within a pre-determined sum up to the tender stage.

- It attempts to measure and value the cost of work after it is designed and *before* it is executed (built).

- Is a process of pre-costing the design proposal which attempts to represent the total picture of anticipated costs in a way which provides a clear statement of issues and isolated the courses of action and their relative cost, so as to provide a guide to decision-making.

- Cost planning aims at ascertaining costs *before* many of the decisions are made relating to the design of a building. It provides a statement of the main issues, clarifies the various courses of action, determines the cost implications of each course and provides a comprehensive economic picture of the whole.

- As long ago as 1959 (RICS Cost Planning Report) the emphasis was on viewing cost in balance with other design criteria:

    A systematic application of cost criteria to the design process, so as to maintain in the first place a sensible and economic relation between cost, quality, utility and appearance, and in the second place, such overall control of proposed expenditure as circumstances might dictate.

    Grafton, 1961 (NP)

Cost plan

The document that states how much is to be spent on each functional element of a proposed building in relation to a defined standard of space and quality.

Element

Its full title is Functional Element, but is now more commonly known as an Element. The RICS (1969: 1) describe it thus:

…a component that fulfils a specific function or functions irrespective of its design, specification or construction.

Morton and Jaggar (1995: 41) describe it as, '… that part of a building which performs a specific function independently of quantity or quality'

Examples of such (functional) elements are, external walls, roof and upper floors. The RICS (1969) Building Cost Information Service (BCIS) Standard Form of Cost Analysis lists 33 standard elements (with 6 group elements), whilst the AIQS (2000) has defined 47 Standard List of elements (with 7 group elements) in the *Australian Cost Management Manual: Volume 1*.

Cost plans are prepared in a standard format using the RICS (1969) BCIS Standard Elements.

Cost Analysis   A systematic breakdown of tender cost data into the standard elements of a building. Bills of quantities aided the cost analysis activity. However, the reduction in the use of bills of quantities in contracts and procurement methods over recent years has meant that we rely more on the analyses in the published price books. Cost analyses facilitate the comparison between projects and the development of recording cost data for future cost planning. Costs are often expressed on a unit basis such as costs per square metre of gross floor area, or element quantity.

The Appendix contains a copy of The Royal Institution of Chartered Surveyors (RICS) (1969) and BCIS *Standard Form of Cost Analysis: Principles, Instructions and Definitions*. This document provides the authoritative guide to terms used in cost planning and cost analysis of projects. See the Appendix for details.

Cost Checking   The process carried out by the cost planner during the Final Proposals and Production Information stages to ensure that neither the elemental cost targets nor the overall *cost limit* is exceeded.

Cost Control   The term embracing all methods of controlling the cost of a building from the inception stage through to the completion and preparation of the final account, handover and into occupation of the building through life cycle costing and facilities management. Cost control is relevant at all these stages in the development and use of buildings.

Note that in this text we are only concerned with the cost control processes and techniques used up to the tender stage (pre-construction).

Cost limit   The total projected cost of the project in the *cost plan*. It is the sum of all the *elemental cost targets*.

Cost Management	This is a broader term than cost planning as it embraces the total development of a project and the use of techniques to manage the cost from the inception to completion of the project including the life cycle costs into the building's operation (see AIQS, 2000: 2–8).
Cost Target	An allowance included in the *cost plan* for each *element*. It is *not* a fixed figure and can be adjusted to take account of specification, quality and area changes as the design progresses. Adjustments in one elemental cost target must be reciprocated by compensating changes in one or more elements to maintain the overall *cost limit*.

AIMS OF COST PLANNING

As noted in earlier chapters we must never forget the aims and objectives of cost planning. In essence the purpose of cost planning is to:

- Ensure that clients are provided with value for money.
- Make clients and designers aware of the cost consequences of their proposals.
- Provide advice to designers that enables them to arrive at practical and balanced designs within budget.
- Integrate costs with time and quality.
- To keep expenditure within the amount, or limit, allowed by the client.

Question 4.1
Staying Within Budget

This chapter is mainly concerned with the third aim noted above; keeping expenditure within the amount allowed by the client. Although in achieving this aim the other two aims can also be achieved.

In keeping expenditure within the amount allowed by the client, what does this mean?

Does it mean that the final account should not exceed the first estimate?

Or, does it mean that the final account can exceed the first estimate?

PRINCIPLES OF COST CONTROL

Before we become involved in describing the process of cost planning and cost control through the various stages, a few comments are needed to give the basis for cost control.

Cost control belongs to a family of control systems that are based on the following three basic principles:

- there must be a *frame of reference* (cost limit);
- there must be a *method of checking* (cost target);
- there must be a *means of remedial action* (cost check).

Frame of Reference

Having a frame of reference in a cost control system consists of two stages:

- establishing a realistic first estimate;
- planning how this estimate should be spent among the *elements* of the building.

When the client has accepted the realistic first estimate, this sum is considered as the *cost limit* for the project. By the cost limit, we mean the cost agreed between the client and the design team as the upper limit beyond which the project may not be pursued.

The second part is to establish how the estimate is to be spent on the various parts, or elements, of a building. This entails dividing the total cost limit into a number of element cost limits called cost targets with one cost target for each element of the building.

The standard list of elements commonly used in the UK that published by the RICS (1969) in its *Standard Form of Cost Analysis* (see Appendix). The equivalent list used in Australia is published by the AIQS (2000) in its Australian *Cost Management Manual: Volume 1*.

Method of Checking

Cost targets are checked as the design advances. The cost planner, in conjunction with all members of the design team, must detect and measure variances from the cost targets previously established. Accordingly, the cost limit (or estimate) is checked (or spent) as it was originally established.

The name given to this activity is cost checking. The check is carried out using an appropriate technique with the information available at the particular stage for the specific element. Variances between the sum total of the cost targets and the overall cost limit will require remedial action to be taken.

Remedial Action

Remedial action should be taken before any serious damage is done or work is so advanced that a great deal of abortive work is the result. That is, before the decisions creating over-expenditure are irrevocably incorporated into the project. The remedial action must be taken quickly and, ideally, as soon as the cost overruns become known and the decisions communicated to everyone in the team. Remedial action thus ensures that total project expenditure is contained within the *cost limit.*

Occasionally, suitable remedial action is impossible, and additional finances must be obtained to complete the project. The client will generally be requested to supply the additional funds. This implies that the initial estimate was incorrect and will certainly embarrass the design team. However, the presentation to the client of a carefully completed and detailed revised budget is the most convincing argument to justify the need for additional funds. The alternative does not bear consideration.

The three principles of project cost control are summarised in Figure 4.2.

Figure 4.2 summarises the three principles of cost control in diagrammatic form. The cost limit (budget) is established during the earliest stages of a project and provides the frame of reference for the total cost planning process. The cost limit is divided into specific and separate cost targets, which may be group elements to begin with, but as the design develops they will be based on the individual elements and eventually in some cases down to sub-elemental level. This division of the budget most commonly takes place during the Detailed Proposals stage, but it can take place at the earlier stage of Outline Proposals.

These targets are periodically checked as the design progresses and remedial action is taken as required to bring the predicted costs within the overall cost limit. The original targets may be adjusted from their first targeted figure, but the overall cost limit should be maintained. This should only change if there is no other way of absorbing any increases in the original cost limit. Most of the cost checking and remedial action takes place in the latter two stages of Final Proposals and Production Information, as the tender documentation is being produced and finalised.

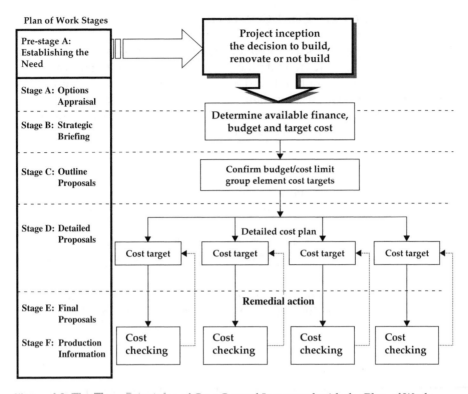

Figure 4.2 The Three Principles of Cost Control Integrated with the Plan of Work
Source: Adapted from Smith and Love (2000: 11)

Question 4.2
Cost Targets

When the client has accepted the realistic first target cost, this sum is considered as the *cost limit* or budget for the project. By cost limit, we mean the cost agreed between client and design team as the amount beyond which the project may not be pursued. The next process is establishing how the cost limit is to be spent among the elements of the building and this entails splitting the cost targets into a number of smaller cost limits, one for each element or group of the building.

Should the sum of the small cost limits (usually known as cost targets) equal the cost limit of the building project?

The first principle of a cost control system, creating a frame of reference, when applied to a cost control system, consists of two stages. What are the two stages?

PRE-CONSTRUCTION COST CONTROL

The RIBA (2000) describes the inception and pre-construction work stages in the Plan of Work for a traditional procurement process. These stages also provide the key stages for building design cost planning process. These stages were illustrated earlier in Table 2.2 and this table is shown again slightly modified in Table 4.2. The primary cost planning activities noted earlier are also summarised in this table.

Table 4.2 RIBA Plan of Work Stages

Inception or Feasibility			Pre-construction Stages			
Pre-stage A	Work Stage A	Work Stage B	Work Stage C	Work Stage D	Work Stage E	Work Stage F
(+1)	1	2	3	4	5	6
Establishing the Need	Options Appraisal	Strategic Briefing	Outline Proposals	Detailed Proposals	Final Proposals	Production Information
Cost Planning Activities at Each Stage						
Establish the budget	Cost of preferred solution	Target cost	Prepare initial cost plan	Firm cost plan	Cost checks, design against cost plan	Final cost checks of design against cost plan

The RIBA (2000) identifies six plus one stages, making seven stages in all up to the tender stage.

Although these stages can be defined and described as being distinct, in practice it may be more difficult to discern the exact conclusion of one stage and the commencement of the next. The development of information for a project often proceeds in a disjointed fashion and commonly parts may be developed to a high level of detail while others may be rudimentary up to, and sometimes into, the tender stage. It is useful, nevertheless, for the design team to appreciate the overall framework (or model) of operations to guide their activities. The cost planner must recognise that despite these practical difficulties he or she should guide the decision-making to those design decisions that have significant cost effects and were discussed in Chapter 3.

Each of these stages will now be reviewed.

Pre-stage A: Establishing the Need

Project Inception or Initiation

The client has to establish the project's objectives and identify the limit of the costs that are reasonable in the circumstances. The organisation, faced with a problem of expansion, contraction, changing markets, greater competition and major legislative requirements for its facilities would address the situation by preparing an outline business case (RIBA, 2000: 11). In fact, Gray and Hughes (2001) term this preliminary stage as the *Business Case*.

The business case should follow an investigation into the nature of the problem and the real need that has to be satisfied. The organisation may need to establish a working party or reference group to ascertain the need for the building and develop a statement of need. Gray and Hughes (2001: 93–94) have developed a checklist for information needed for the statement of need. A statement of need will form the basis for determining whether a building is needed.

The resulting business case arising from the statement of need with input from all significant parties on a working party or review group such as users, key stakeholders, project manager or project sponsor and an architect or other building adviser. The final business case should describe the options available for solving the problem (including no-build solutions), outline benefits of each option, likely costs and return on capital investment and an initial assessment of risk for each alternative strategy.

So, in summary for this first stage the client has to establish a representative client management team, appoint a project manager or project sponsor, possibly appoint a cost consultant and this group should identify the objectives (or desires or aspirations) and physical scope of project, the standard of the quality of the building(s) and services (if necessary), advise on the time frame and establish a realistic first *budget*.

Stage A: Options Appraisal

The RIBA (2000: 11) describe this stage as,

> Identification of client's requirements and of possible constraints on development. Preparation of studies to enable the client to decide whether to proceed and to select the probable procurement method.

The project manager and the client may see the Appraisal stage as a 'reality check'. The statements made in the business case and the statement of need will now be subject to their first scrutiny to test whether they are realistic, affordable and achievable within the constraints that are known at this early stage. A number of basic aspects will have to be reviewed by the client, project manager and the emerging design team:

- functional performance (fitness for purpose);
- quality, profile, image (firmness and delight);

- timing for operational implementation (the programme);
- available finance or funding (cost limit or budget);
- value for money.

The most critical part of this stage is the development of realistic options that adequately engage the client's problem environment and satisfy the project objectives. The client and the project manager, lead consultant or a facilitator, may organise a workshop involving a number of participants who have knowledge of the client's operations, problem environment and a need to change. These participatory workshops may be organised formally in a strategic value management structure (Dallas, 2006) or using the more open *strategic needs analysis* methods (Smith, 2005). The main focus of these workshops is to understand the client's problem and to identify a range of solutions, strategies or options for consideration and review by the group. The number of options developed may be considerable and the workshop facilitator must guide the participants into a manageable number of options for review and further analysis.

The workshop participants should decide on the common criteria for evaluation of the options and these should match the client's objectives. Assistance in the development of the options and their evaluation may be gained from a range of techniques now being used in the options development and appraisal stage. Again, value management has an approach that can be readily used (Connaughton and Green, 1996), Quality Function Deployment (QFD) has also been used successfully (Gray and Hughes, 2001: Akao, 1990), *Expert Choice* (Analytic Hierarchy Process) by Saaty (1990a, b) and several applications using *Situation Structuring* (Dickey, 1995; Smith and Love, 2004). This review and evaluation may cause some options to be discarded and others merged into a single option. A do-nothing or status quo option should always be included to test desire for change or measure inertia in the stakeholder group. The options during this stage need not all be building projects and all feasible options to 'solve' the problem need to be considered.

If the chosen option (or options) requires a fully developed building project or contains part building, then information on the site, existing building (if any), location, accessibility, functional effectiveness, and financial implications of each option. The cost guidance offered at this stage will be limited and can only match the level of detail and type of information available for each option. Where the option has a building content then the cost adviser would have to ensure that the basic data for costing the option, such as the overall nature of the project, units accommodated or amount of gross floor area needed would be estimated and advised. If much of this information is not available then a lump sum costing based on a similar project, or some order of costs may have to be used to provide a financial dimension to reviewing the options. Of necessity, the costs would have to be heavily qualified in terms of their accuracy and use. The participants must be made fully aware of the scope for error at this stage would be extremely high.

The AQUA group (2003: 45) summarise the cost activities for each option that has a built facilities component or whole building(s) at this stage as consisting of:

- cost aspects for site appraisal studies;
- cost aspects of design studies;
- initial capital cost studies;
- value management and cost benefit studies;
- cost of preferred solution(s).

The conclusion of this stage may result in a single option or possibly more than one option in some cases and the expectation is that further evaluation work is required before a final decision on the preferred option can be taken. The basis of the Options Appraisal will vary with project environment and should be left to the discretion of the client, facilitator and the participants to decide on its format in the final report. This report then becomes the basis of the next stage in the process, Strategic Briefing.

Stage B: Strategic Briefing

The RIBA (2000: 14) describe this stage as,

> Preparation of Strategic Brief by or on behalf of the client confirming key requirements and constraints. Identification of procedures, organisational structure and range of consultants and others to be engaged for the project.

The early Briefing stage has received more attention in recent years because it has been recognised that it is at this stage that a number of critical design and project decisions are made. Not least of which, is whether the project proceeds at all. Thus, more attention has been paid to the processes and techniques available during this decisive stage.

Strategic Briefing is a critical phase in the development of any project. It is a working reference document prepared by the client and it attempts to integrate the business and its implications on the work and operational environment with the mission vision and objectives of the organisation. The strategic brief assists in bridging the communications and professional divide between the organisation and the design team. This brief describes the organisation's strategic management objectives, its future spatial needs and defines the outcomes in strategic or conceptual terms to allow the design team the freedom to explore and innovate to achieve the organisation's objectives. In practical terms, through documenting its strategic management objectives the organisation retains control and influence over the future design investment decisions handed over to the design team. So the design team will have the strategic constraints to its activities clearly defined in this document.

The strategic brief document should be clearly focused on the outcomes of the project for the organisation and its content will address the following issues:

- the organisation's mission/vision statement;
- outline of business case;
- context of project and strategic management decision framework;
- organisation and structure;
- functional expectations of facilities;
- size and capacity expected in new facilities (operational range specified);
- environmental qualities and aims;
- proposed level of servicing (operational and whole life costing);
- building quality levels expected;
- project management structure with key reporting dates and lines of responsibility;
- programme or time frame;
- anticipated procurement path;
- budget or cost limit projected;
- client and user expectations (strategic not prescriptive);
- key performance indicators both during and on completion of project.

The Briefing Stage is one involving the two component activities highlighted in the table above. These are inception and feasibility. The Briefing stage has received more attention in recent years because it has been recognised that it is at this stage that a number of critical design and project decisions are made. Not least of which, is whether the project proceeds at all. Thus, more attention has been paid to the processes and techniques available during this decisive stage.

The AQUA Group (2003: 45) have identified three major cost advice activities during this stage of the process:

- Whole life costing
- Target Cost
- Cash flow constraints.

The NPWC (1989: 2) in reviewing the equivalent stage in the pre-construction cost management process has identified the aim of this stage to provide:

> A first cost indication to the client based on an outline statement of the client's needs. The indicative cost is intended as a guide for feasibility and planning purposes, it is not an estimate and may not be quoted as such.

The objective of this stage is to test the feasibility of the project:

> To establish an initial budget for client consideration or to confirm that the client's budget is feasible (AIQS, 2000: 2–12).

An important function of this stage is to test the feasibility of the chosen option or the options still being considered by the client and design team. As noted in the contents of the strategic brief the design team should provide the client with an appraisal and a recommendation so that a decision may be made on whether, and in what form, the project is to proceed. This is a crucial stage in the life of the project.

Cost planning begins formally with the preparation of the client's initial brief.

An effective cost control system must then be set up which initially incorporates the first principle of cost control:

- establish a realistic first estimate (cost limit);
- decide how to allocate this estimate among the elements (cost targets).

The cost planner and the design team must consider the balance between the client's requirements in the brief, and the amount of project funding required to fulfil these needs and requirements. If the client has not provided a cost limit for the project then the design team should produce its own estimate which, when accepted by the client, effectively becomes the cost limit or budget. On the other hand, the design team may have to confirm that a building can be completed within the client's previously stated cost limit.

In these early stages of a project, the client will either be in a position to agree a budget or will need to reconsider the project content. The design team and the client may have to repeat this procedure several times before a sensible balance between brief and budget can be achieved. Changes made to the budget at this stage and during the later stages must involve the client in the decision and the authorised changes must be properly recorded and communicated by the project manager.

These early stages are likely to involve tentative design and organisation changes and they should be made with an awareness of the capital, life cycle, operational costs and revenue implications. The financial commitment caused by any alteration to the brief must be made explicit by the design team and involve the client team fully, so they can model the effects on the client organisation and operations.

At the feasibility stage the information available to the design team is often related to the amount of floor space on the project (gross floor area), the standard of accommodation provided (budget to prestige), and the functions (uses) for which the building is intended. Equating total costs of the project with the expected revenue or income from the proposed project is an important appraisal to be made at the financial feasibility stage. Building or capital costs form a significant part of the total cost side of the developer's equation (along with land, finance, profit or return, fees and marketing).

The various methods of development appraisal that are used during the Feasibility Stage are not considered in this text, but more detail of these methods can be found in the property literature such as Robinson (1989), Cadman and Topping (1995), Isaac (1996, 1998), Brown and Matysiak (2000).

Flanagan and Tate (1997: 227–241) show an example of a Feasibility stage cost plan, which provides a good practice template for the cost planning activities at this Strategic Briefing stage.

This first stage is important because it establishes the brief and sets the cost limit or budget for the project. We stress that the client's objectives are the foremost consideration and that the aim of cost planning is directed towards their achievement. Unless the client has imposed a cost limit before the briefing process commences, it is the primary function of the early stage cost planning procedures to establish a realistic budget. When the budget and the brief are agreed (and they may change throughout the subsequent stages) they provide the *frame of reference* within which design and organisational decisions are considered.

Stage C: Outline Proposals

The RIBA (2000: 4) summarises the main activities at this stage as being:

> Commence development of Strategic Brief into full Project Brief. Preparation of Outline Proposals and estimate of cost. Review of procurement route.

The development of the project intensifies and the cost and economic aspects receive more attention as the project information expands. The AQUA Group (2003: 45) have identified four major cost advice activities during this stage of the process:

1. assessment of economic constraints;

2. cost studies of design and energy options;

3. initial cost plan (with group element cost breakdown);

4. cash flow projections.

NPWC (1989: 2) describes this stage in these terms:

> The Outline Proposal Cost (Preliminary Estimate) ... is the first cost estimate and is usually prepared on the basis of the client's brief, investigated site conditions and preliminary sketches ...at this stage of the design development, it may be necessary for the consultant to evaluate the comparative cost of various outline proposals.

The objective or purpose is described by the AIQS (2000: 2–12) as:

> To identify the best means of satisfying the requirements of the Brief, the resultant scheme becoming the basis for the sketch design.

Each of the alternative outline proposals are investigated with alternative building configurations being explored. For instance, low rise versus multi-storey buildings (AIQS, 2000). The brief is developed further, but there is still insufficient detail and information for individual cost targets to be prepared for each separate element listed on the RICS (1969) *Standard Form of Cost Analysis* (see Appendix). The most realistic breakdown of the budget for this stage would be to the broader subdivision of the eight group elements as listed below:

1. Substructure

2. Superstructure

3. Finishes

4. Fittings and Furnishings

5. Services

6. External Work

7. Preliminaries

8. Contingencies

Commonly, the Group Elements costs from the RICS (1969) list may be estimated using 'notional' or standard breakdowns for similar types of building as a starting point for the cost targets. The costs of these group elements are calculated by studying the cost allocations for these group elements in similar buildings. The origin for such information can be found from internal sources such as past projects in the quantity surveyor's/architect's offices and from external sources such as the BCIS or building price books.

Inspection of BCIS cost analysis records of similar buildings would provide the cost planner with the starting point for breaking the total budget into similar group element targets based on the chosen analysis. The content of the BCIS is discussed in more detail in a later chapter. Another related approach using the BCIS cost database would be to take a few similar buildings and average the percentage costs of the group elements for the new project. However, this process needs to be carried out with care as individual projects may disturb this notional or typical cost distribution.

In other countries, for instance in Australia, where the BCIS is not used the cost planner would have to use publications such as the following:

- *Rawlinsons* (2006) (annual publication with quarterly updates).
- *Cordell's* Building Cost Guides (quarterly publications).
- *The Building Economist* (quarterly publication by the AIQS).
- Various publications by firms of professional quantity surveyors or building cost consultants (e.g., Rider Hunt and WT Partnership).

At this stage, because of the lack of detailed information, all that may be accomplished is a rudimentary subdivision of the budget into reasonable percentages or proportions based upon similar past project(s). The essence of the exercise is not absolute accuracy in the pricing of the group elements, but a sensible subdivision of the budget based on successful, balanced similar past projects.

For example, in practice, this may consist of a single-storey primary school with a budget of £1.5 million having the following percentages allocated to each group element.

Proposed Primary School

Group element	Allocation (%)	Target (£)
1. Substructure	8	120,000
2. Superstructure	25	375,000
3. Finishes	12	180,000
4. Fittings	5	75,000
5. Services	15	225,000
6. External Work	15	225,000
7. Preliminaries	12	180,000
8. Contingencies	8	120,000
Budget		£1,500,000

Note: This is not a real project!

Smith and Love (2000: 35–54) show the equivalent *Outline Proposals* stage cost plan in Australia using the AIQS model and local sources of information as basis for developing the cost plan with its group element and elemental cost breakdown. Flanagan and Tate (1997: 240–250) show an example from UK practice.

The Property Services Agency (1981: 9, 10) *Cost Planning and Computers*, lists the most important characteristics that ought to be considered at this stage as being:

- type of project;
- size and general arrangement of project;
- quality;
- detail design;
- complexity;
- construction form or system;
- contract arrangements;
- location;
- occupation date required by client;
- inflation;
- degree of repetition within the project;

- the architects and engineers responsible for design;
- degree of innovation within the project;
- weather or ground conditions;
- market conditions;
- quality of project information.

These characteristics create costs and they interact with each other to create the total cost of the project. In recent years some cost planners have created cost models and expert systems, which attempt to allocate the major cost variables to each other and to predict the total cost of a project. Though the method of calculating or predicting the total cost of the project and its group element breakdown seems crude and simplistic, it does nevertheless, provide a robust starting point for cost planning. This is because the initial cost subdivision is soundly based on the cost analysis of a successfully completed and analysed project from the BCIS cost database, which has a practical balance of expenditure between the various targets (elements).

This genesis of the cost planning process requires great skill in applying judgement and expertise. Although a similar past project, or projects, have been chosen to guide the cost allocations in the new project, the allocations must be adjusted to take account of obvious and significant differences such as:

- functional content and differences;
- user capacity and scale;
- obvious design divergences (single versus multi-storey);
- construction methods/specification;
- market conditions (boom versus bust);
- time (past versus present versus future values);
- unusual/difficult locations;
- other special factors depending on individual project circumstances.

Once these broad cost allocations are determined as indicated earlier this becomes the Outline Cost Plan. The client is presented with this as a realistic first estimate (if none has been given to date) or as a confirmation of the cost limit agreed at the earlier Options Appraisal and Strategic Briefing (Feasibility) stages.

Depending on the client, and the type of project, it may also be necessary to confine the cost information at this stage to a range of costs (rather than a single budget figure). The preparation of a firm estimate can then be made when the Detailed Proposals (Sketch or Scheme Designs) have been developed.

The end of this stage should provide a solid basis for the further and continuing development of the project in detail.

Question 4.3
Briefing Stage

Many authors have commented that the briefing stage is the most important stage in the whole construction process. Do you agree? Support your view with an argument for or against. Try to use some actual experiences to support your answer.

COST PLANNING THE DESIGN

Stage D: Detailed Proposals

The completion of the Outline Proposals stage provides a significant milestone in the development of a project. Many of the important decisions affecting the cost/time/quality of a project will have been made and the design now progresses within the parameters or constraints established during the briefing stage.

The detailed design stage commences the process of putting 'flesh-on-the-bones' of the project and the cost plan becomes the centrepiece of the design and its relationship with the budget .

The AQUA Group (2003: 45) have identified six major economic and financial cost advice activities during this stage of the process:

- further cost studies;
- elemental cost plan (divided into detailed element costs);
- updates of cost plan as design develops;
- cost effects of compliance with statutory requirements;
- firm cost plan;
- cash flow projections.

Referring to the AIQS (2000: 2–13) provides a guide to the activities at this stage, which has a different title to the RIBA Plan of Work. It is known as the Sketch Design stage as distinct from the Detailed Design stage and its purpose is:

> ... to ensure that the overall design is the most effective available in terms of the requirements, to confirm or set the final budget and to establish elemental cost benchmarks.

The objective (NPWC, 1989: 2,3) is:

> ... to ensure that the overall design is the most cost effective in satisfying the requirements of the brief; to confirm or set the final budget; and to establish elemental cost targets.

The design team completes and agrees user studies with the client and other stakeholders and these are used to ensure that the design concept complies with the developing brief. The design team will agree and complete the project brief. The major activity is to make progress towards the full design of the project, including planning arrangement, appearance, constructional method, outline specification and cost. Not all the layouts will be completed. Some further work may be required in the next stage, but a substantial amount of the critical work will be finished during this stage. For typical details of drawings and a standard cost plan for this stage see Smith and Love (2000: 55–90 and 149–157 in Appendix A).

Additionally at this stage, all necessary planning approvals are obtained. These include planning approvals from local, regional and other planning authorities. If these authorities request any changes then these have to be incorporated in a revised design and cost plan.

Naturally, during this stage, sketch plans are produced and the cost plan formulated. This second cost plan is a more detailed statement of how the design team proposes to distribute the available money among the elements of the building. Whereas the Outline Proposal Cost (see earlier) distributed the budget to the group element level of detail, the Detailed Proposals cost plan distributes the budget to the individual element level of detail, wherever possible.

Sketch plans and layouts are produced concurrently with the detailed cost plan. The sketch plans at this stage are more detailed than at the Outline Proposals stage. They will show the location of all walls, windows, doors, partitions, stairs, sanitary fittings, etc. and contain details of elevations and heights.

The project drawing now contains more information about the size, or extent, and type of functional spaces to be provided. The internal and external building services are also outlined. Significant dimensions and heights are clearly shown on the plan and related elevations of the building. Information about the site, including the type of sub-soil is given as a note and the elevations and sections enable the cost planner to measure some of the major elements reasonably accurately. The information data base for the project is rapidly expanding to provide the means for more accurate cost plans to be produced and to check whether the initial cost targets established were realistic.

In preparing the cost plan it must be borne in mind that at this stage, only total areas and numbers are usually known, together with an indication of quality. At this stage the design team cannot be expected to enter into detailed consideration of specification of every element. This is achieved by using the more developed information of this stage. The Detailed Design Stage cost plan is based on the following information:

- complete project brief;
- outline cost plan (from previous stage);
- final dimensioned sketch plans;
- elevations and sections;
- structural sketches with dimensions of members indicated;
- schedule of finishes;
- site layout;
- specification notes.

When building approval is sought from the planning/building authority most of this information listed above has to be provided to gain the necessary approval.

All participants in this stage (client, project manager, architect, quantity surveyor, structural engineer, services engineer) will have participated in the development and completion of the project brief. At the end of this stage all these participants must have signalled their agreement with the completion of the project brief by signing off its completion.

Question 4.4
Cost Plan or Estimate?

The cost plan provides a frame of reference and this is established during the sketch plan stage. It is not an estimate, but a working document for action.

How does a traditional estimate differ from a cost plan?

It is essential that the production of a cost plan is seen as a joint enterprise by the whole design team. In the design of a building project the skills of each design team member are called upon, and each member can be seen to be responsible for incurring expenditure. This liaison and co-operation between the members of the design team helps to ensure that the standards set are achievable within the cost limit.

Stage E: Final Proposals

Using the RIBA (2000) Plan of Work as a guide the major activities at this stage can be summarised as:

- sanction and complete final layouts.
- co-ordinate all components and elements of the project.
- cost checks design against cost plan.
- update cost plan.
- final decision on procurement method.
- review design and cost plan.
- update cash flow projection
- prepare submission for statutory approvals.
- design (final proposals) is now frozen.

The aim of this stage is to convert the detailed design into the finally agreed proposals that should not be substantially changed from this point onwards. The other important aim of this stage is to have the entire necessary documentation ready for submission to the various planning and other statutory authorities and to gain their approval for the project content. This is a critical stage in the life of the project and it demands a great deal of attention to detail by all the members of the design team.

Gray and Hughes (2001) consider that this stage is where all the major systems of the project are specified, detailed and integrated into the final project proposal. Much of the design and engineering team activity is focused on evaluating alternative technological solutions before making a submission to gain detailed planning approval. Failure to gain this planning approval will delay the project and may cause serious financial, design and functional disruption. So, this is a critical stage in the life cycle of the project.

The completion of this stage should see the following client, design team and authorities' approvals for:

- The Final Proposals documentation match the priorities and statements contained in the project brief.

- The elemental cost plan is within budget and is based on the agreed and approved Final Proposals documentation.

- Detailed planning approval is achieved.

- All parties sign off the project and the design is now frozen.

For an example of a cost plan at the end of the Detailed Proposals and Final Proposals stage refer to Flanagan and Tate (1997: 254–282).

With the design now fixed and approved it moves into the final design stage before tenders are called, that is, Production Information.

Stage F: Production Information

The design team will be fully occupied at this stage converting the Final Proposal information into the Production Information that becomes the tender documentation for the project. The RIBA (2000) envisage the main activities at this stage being centred on:

- Prepare all co-ordinated Production Information including location, assembly and component drawings, schedules and specifications as recommended by Co-ordinating Committee for Project Information (1987) and the Building Project Information Committee (1987).

- Applications for statutory approvals completed, including building regulation approval.

- Provide all project information for final cost checks of design against cost plan.

The AIQS (2000) go straight to the financial heart of the matter and succinctly summarise this stage as:

> … to ensure that the overall detailed design is contained within the budget.

Within this terse general objective is contained a great deal of cost monitoring and checking activity on the part of the whole design team.

The NPWC (1989: 4) summarises the responsibilities of the cost planner as follows:

- carrying out all necessary cost checks on each element as the design details are developing;

- comparing cost checks with the cost plan (prepared in the sketch plans stage) and progressively reporting on variances by using reconciliation statement;

- ensuring that the design team is aware of the cost implications of its developing drawings and specifications so that the completed design is contained within the agreed cost limit;

- closely liaising with the project team at all times so as to eliminate any abortive design time; and

- identifying and advising the team of any possible cost savings to the project.

In the earlier stages the cost planning activities have been mainly concerned with planning what should be done with the money available for a building. At the Production Information stage the emphasis changes to *action*. At this stage final decisions must be made on every matter related to design specification, construction and cost. Every part and component of the building must be fully designed and all designs must be completely cost checked, and remedial action taken if necessary.

Detailed designs are prepared for each element of the building. All designs must be complete, construction problems solved, and material, sectional details, fixings, finishes all decided.

The drawings prepared by the design team – architect, engineer(s) and consultants – do not have to be the final version of the working drawings that will be used on site. Pencilled dimensioned sketches are sufficient, provided all design solutions are complete. The presence of detailed drawings and specifications at this stage permits the cost planner to prepare accurate measurements and relatively detailed quantities based upon a technique called the approximate quantities method of estimating. (This is described in detail in later chapters). The cost planner measures approximate quantities from the detailed designs for each element in turn, and prices these quantities using traditional estimating methods. The cost planner then informs other members of the design team of the estimates for each element using the RICS (1969) Standard List of Elements and format.

Smith and Love (2000) have described this stage as one where a great deal of cost monitoring activity takes place as the working or production drawings, specification, schedules and other project information produced. Cost planning activities must be integrated into the documentation process and as the information is being developed and produced the cost planner is consulted so the cost checking activities can be completed at the earliest stage. This is likely to reduce the possibility of major changes disrupting work in other elements and parts of the project.

If the budget is likely to be, or is, exceeded as a result of the production of the tender documents then remedial action must be taken. Due to the interrelated nature of elements and characteristics of a project it is essential to involve and inform all members of the design team about these decisions and proposed actions. Regular meetings to communicate information are essential in this process and any agreed action must be promptly and clearly documented and distributed around the team.

Flanagan and Tate (1997: 283–300) provide details of cost checking activities carried out during this stage, the equivalent of their Detailed Design from the previous RIBA Plan of Work. An example of a detailed cost check of a whole project is also given in NPWC (1989: 11–16), Appendix B and in Smith and Love (2000: 91–147 and 159–170).

If the cost checking is not carried out thoroughly and comprehensively then all the previous cost planning and design work will have been wasted.

When cost checking reveals that remedial action is necessary, two courses of action are necessary:

- If the cost of the element design is greater than a realistic cost target, then the element design should be changed so that it is within the cost target.

Alternatively,

- If the cost check of the element design shows that the cost target is unrealistically low, cost targets should be adjusted downwards throughout the cost plan, thereby releasing surplus funds for the element in difficulties.

In carrying out this adjustment, the cost planner will not alter the overall project cost limit. This is the whole point of cost planning – keeping expenditure within the amount allowed by the client. The cost plan can be changed even at this stage, as long as the cost limit is not altered. If the cost of the element design is just within the cost target, no remedial action will be necessary, but the cost planner should inform the rest of the design team. The production or documentation of the final working or tender document drawings for that element can then proceed.

If the cost of an element is significantly below the target cost, then surplus funds may be released for other elements as required.

Each element is analysed in turn and any necessary action is taken and implemented. Care must be taken where a decision for change in one element will affect the design choices in another element, or series of elements. The full effect of a change must be carefully traced through the whole design and not be artificially compartmented for simplicity and administrative convenience. Through experience it is known that certain elements are more likely to give trouble than others, so the cost checks on these elements should be given the highest priority.

The working or tender documentation drawings provide detailed information to the cost planner about the choice or specification of materials and details of the important parts of the construction. The design team has made decisions on the type, size or thickness of floor finishes, windows, internal and external walls, internal finishes, roof construction and coverings. These drawings would be read in conjunction with a specification document and any schedules that have been prepared. These documents may only be in draft form, but they will provide a suitable basis for pricing the particular element in detail. The cost planner can prepare accurate descriptions of the materials and make detailed measurements for each element that will provide a sound basis for forecasting the total cost of the project before it goes out to tender. However, the main purpose of this stage is to conduct detailed cost checks for each element to ensure the cost targets and the cost limit have not been exceeded. If they have, then remedial action must be taken which will involve changes to the material and possibly the extent of some elements. The application of these changes to the design will be discussed in the later chapters. Similarly, excessive savings in an element or group of elements may indicate a lack of expenditure that could result in poor standards. This should also be considered before the cost of those elements is confirmed.

Once these elements have been confirmed these preliminary detail drawings become the working or production drawings that become the basis for the tender documentation drawings and from which the bills of quantities are prepared if that is part of the contract documentation.

A danger of conducting cost checks in isolation and of aiming for low capital costs is that a building of low quality may be produced. If the result of these actions is a low quality building, then it may be appropriate for the design team to approach the client to confirm whether this is the intention. If not, then the client should release additional funds to improve the overall quality of the building. In this situation the cost plan is a powerful tool for arguing the case for more funds. However, it is not desirable for this to occur at this late stage in the process.

Once all the element designs have been cost checked, a final cost check is done. This ensures that nothing has been overlooked. If a final cost check is not carried out and as a result one or more elements are overlooked, then the cost limit may be over-spent. That is, the main purpose of cost control will not have been achieved. The final cost check is, therefore, an important part of the cost control procedure at the documentation (detail design) stage.

The flow of design team activities at this stage is summarised below in Figure 4.3. At the Production Information stage, principle two (there must be a method of checking) and principle three (there must be a means of remedial action) are incorporated into the cost control system. These are illustrated within Figure 4.3.

If the final cost check confirms that the project is financially sound, design team activities can proceed into the Tender Documentation Stage; working drawings, schedules, specifications, bills of quantities, contracts, tenders. The project can then go to tender, which is confirmation of whether all the cost planning activity described in this chapter has achieved its objective.

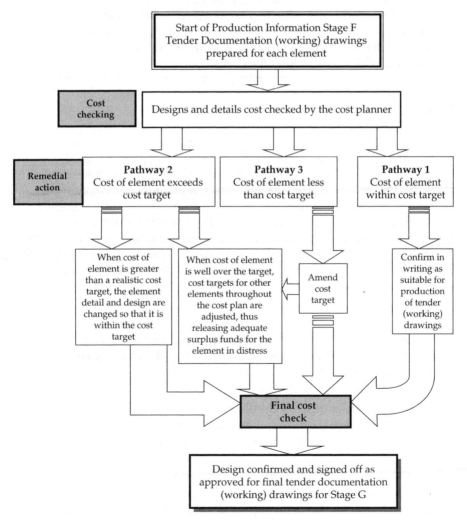

Figure 4.3 Pathways to Final Cost Checking
Source: Adapted from the Ministry of Public Buildings and Works (1968: 94).

A final pre-tender estimate based on this tender documentation can be prepared and if the contract requires bills of quantities to be prepared from this completed documentation then the pre-tender estimate will consist of pricing the final bills of quantities. If quantities do not form part of the contract then the cost planner can prepare the tender cost plan from the tender documentation.

Question 4.5
Cost Checking

The cost plan is used continuously throughout the tender documentation stage as a method of checking that the detailed design is contained within the cost limit. If remedial action taken as a result of this cost checking were to produce a building of low quality, would the design team be justified in continuing without consulting the client?

Chapter Review

This chapter provides details of the essential principles and framework to cost planning. The need for cost planning in building project activities is the same for any economic activity in a society such as ours. It is absolutely essential. We stressed the theoretical underpinning for understanding the nature and purpose of cost control activities. More importantly, it is essential for the future cost planner to understand these principles so that when pressures for change do occur the basic principles are rigorously maintained. The role of the RIBA and the BCIS is in providing a standard approach to the Plan of Work with a cost reporting structure that aims to achieve cost control and management.

The context of the process of cost planning and cost control was given in an overview of the whole process of pre-construction or pre-contract cost management. The various stages identified in the RIBA Plan of Work were identified and described.

You should be able to describe the aims of cost planning; realise that the incessant demands for cost control on our building projects will never abate; and understand the principles to be adopted by any cost control system in practice.

The nature and significance of the briefing stage was emphasised in the entire process of project development. It provides the guiding and controlling cost framework for the project from these early stages into the later detailed design stages, and even into the construction period.

The type, content and extent of information available at each stage must be understood and acknowledged, as it guides and often dictates the types of cost planning technique used. In fact, the development and expansion of information through the various stages of a project's life is a key component in the application of cost planning principles and cost control activities.

Review Question

What are the three principles of cost control?
What are the six (plus one) stages of cost control described in this topic?
Briefly describe the aim of each stage.

References

Akao, Y. (1990) An Introduction to Quality function Deployment in *Quality Function Deployment (QFD): Integrating Customer Requirements into Product Design*, Akao, Y. (editor), Productivity Press, Portland, Oregon, USA, pp. 1–24.

Australian Institute of Quantity Surveyors (AIQS) (2000) *Australian Cost Management Manual*, Vol. 1, Australian Institute of Quantity Surveyors, Canberra, Australia.

Australian Institute of Quantity Surveyors (AIQS) (2001) *Australian Cost Management Manual*, Vol. 2, Australian Institute of Quantity Surveyors, Canberra, Australia.

Brown, G and Matysiak, G (2000) *Real Estate Investment: a Capital Market Approach*, Chapter 7: Valuations and prices, Harlow: Prentice-Hall Financial Times.

Building Economist (quarterly publication), *Journal of the Australian Institute of Quantity Surveyors*, Canberra, Australia.

Building Project Information Committee (1987) *Production Drawings*, National Building Specification Services Ltd, Newcastle.

Cadman, D and Topping, R (1995) *Property Development*, 4th edition, E&FN Spon, London.

Connaughton, J. N. and Green, S. D. (1996) *Value Management in Construction: A Client's Guide*, Construction Industry Research and Information Association, London.

Coordinating Committee for Project Information (1987) *Co-ordinated Project Information for Building Works: A Guide with Examples*, National Building Specification Services Ltd, Newcastle.

Cordell Building Information Services (quarterly) *Commercial/Industrial Building Cost Guide*, Reed Publishing, St Leonards, New South Wales, Australia.

Cordell Building Information Services (quarterly) *Housing Building Cost Guide*, Reed Publishing, St Leonards, New South Wales, Australia.

Dallas, M. (2006) *Value and Risk Management A Guide to Best Practice*, Blackwell Publishing, Oxford.

Dickey, J. W. (1995) *Cyberquest: Conceptual Background and Experiences*, Ablex, Norwood, New Jersey, USA.

Flanagan, R. and Tate, B. (1997) *Cost Control in Building Design*, Blackwell Science Publications, Oxford.

Grafton, P. (1961) Cost Planning Example II in *Cost Planning: Report of Postgraduate Cost Planning Courses*, Lane, R. F. and Lee, F. M. (editors) The London County Council Brixton School of Building and the Royal Institution of Chartered Surveyors, Quantity Surveyors' Committee and the Cost Research Panel, London.

Gray, C. and Hughes, W. (2001) *Building Design Management*, Butterworth-Heinemann, Oxford.

Isaac, D. (1996) *Property Development: Appraisal and Finance*, Palgrave, Basingstoke, UK.

Isaac, D. (1998) *Property Investment*, Macmillan, Basingstoke, UK.

Ministry of Public Buildings and Works (1968) *Cost Control During Building Design*, HMSO, London.

Morton, R. and Jaggar, D. (1995) *Design and Economics of Building*, E & F N Spon, London.

National Public Works Council (NPWC) (1989) *Guidelines for Cost Planning Consultant Services*, Australian Government Publishing Services, Canberra.

Rawlinsons (2006) *Rawlinsons Australian Construction Handbook*, Rawlhouse Publications, Perth, Western Australia. Annual publication.

Royal Institution of British Architects (RIBA) (2000) *The Architect's Plan of Work*, RIBA Publications, London.

Royal Institution of Chartered Surveyors (RICS) (1969) *Building Cost Information Service (BCIS) Standard Form of Cost Analysis: Principles, Instructions and Definitions*, Royal Institution of Chartered Surveyors, London.

Robinson, J. (1989) *Property Valuation and Investment Analysis: A Cash Flow Approach*, Law Book Company, Sydney.

Saaty, T. L. (1990a) *Multi Criteria Decision Making: The Analytic Hierarchy Process*, Vol. 1, AHP Series, RWS Publications, Pittsburgh, USA.

Saaty, T. L. (1990b) *Decision Making for Leaders*, Vol. 2, AHP Series, RWS Publications, Pittsburgh, USA.

Seeley, I. H. (1996) *Building Economics,* 4th edition Macmillan, London.

Smith, J. (2005) An Approach to Developing a Performance Brief at the Project Inception Stage, *Architectural Engineering and Design Management*, **1** (1), pp 3–20.

Smith, J. and Love, P. (2000) *Building Cost Planning in Action*, University of New South Wales Press, Sydney, Australia.

Smith, J. and Love, P. (2004) Stakeholder Management During Project Inception: Strategic Needs Analysis. *ASCE Journal of Architectural Engineering*, **10** (1) pp. 22–33.

The Aqua Group (2003) *Pre-Contract Practice and Contract Administration*, Hackett, M and Robinson, I (editors), Blackwell Publishing, Oxford.

The Property Services Agency (1981), *Cost Planning and Computers*, HMSO, London.

Further Reading

The following texts also provide the necessary background and detail to gain a better understanding of the material presented in this chapter.

Ashworth, A. (2002) *Pre-Contract Studies: Development Economics, Estimating and Tendering*, 2nd edition, Blackwell Publishing, Oxford.

Ashworth, A. (1994) *Cost Studies of Buildings*, 2nd edition, Longman Scientific and Technical, Harlow, England.
     Chapter 10: Cost Planning pp. 197–230.

Ferry, D.J. and Brandon, P.S. (1999) *Cost Planning of Buildings*, 7th edition, Blackwell Science, Oxford.
>   Chapter 15: Cost Planning the Brief, pp. 197–212.
>   Chapter 16: Cost Planning at Scheme Development Stage, pp. 213–236.

Flanagan, R. and Tate, B. (1997) *Cost Control in Building Design*, Blackwell Science, Oxford.
>   Chapter 2: The Purpose of Cost Control, pp. 11–28.
>   Chapter 3: The Principles of Cost Control, pp. 29–42.

Seeley, I.H. (1996) *Building Economics*, Macmillan, London.
>   Chapter 7: Cost Planning Theories and Techniques, pp. 125–143.

The AQUA Group (2003) *Pre-Contract Practice and Contract Administration*, Hackett, M. and Robinson, I. (editors), Blackwell Publishing, Oxford.
>   Chapter 4: Pre-Contract Cost Control, pp. 44–60.

SUGGESTED ANSWERS

Question 4.1: Keeping Expenditure within the Amount Allowed by the Client

- If you answered that this statement is correct then you are right. When a client accepts a first estimate, the first estimate becomes the 'amount allowed by the client'. As long as the client does not order any additional work after he has accepted the first estimate, then the final account should not exceed the first estimate.

  When a client is presented with a final account for £1,000,000 based on an estimate of £800,000, then the aim of cost planning has not been fulfilled.

- If you answered that the final account can exceed the first estimate, then you are wrong. In a straightforward project the final cost of the building should not exceed the first estimate.

  The only possible exception to this is when the client orders additional work after he has accepted the first estimate. The final account may of course be less than the first estimate, but this is a very rare occurrence!

The design team should ensure that cost control continues after the tender stage through to the end of the project. It is not adequate for the design team to relax its efforts after the tender has been received and a contractor appointed and not to pursue sound cost control practices throughout the construction period. Clients are unlikely to be impressed by a design team that abdicates its cost planning and control responsibilities part of the way through the task it has been set.

Question 4.2: Cost Limit and Cost Targets

- Yes, the sum of the cost targets must equal the cost limit. If we take the overall cost limit, and split it into smaller sums of money, then adding together these smaller cost targets must give the overall cost limit.
- The first principle of a cost control system consists of two stages:
  - establishing a realistic first estimate or cost limit;
  - planning how the estimate should be spent among the elements or parts of the building.

A cost target should be established for each element. There are 34 elements in the RICS (1969) standard format (47 elements in the AIQS standard list), so there are likely to be the same (or slightly less) cost targets. Some however, will have a nil value because they are not needed for that particular building type.

In principle two, there must be a method of checking, involving detecting and measuring departures from cost targets and checking that the estimate is being spent as originally established. This is cost checking.

The third principle consists of taking any necessary remedial action. The remedial action should be taken before any damage is done. The action should also be taken within the time allowed for the project. The remedial action thus ensures that expenditure is contained within the amount agreed between the client and the design team; that is, the cost limit.

Occasionally, suitable remedial action is impossible, and additional finances must be obtained to complete the project. The client will be asked to supply the necessary additional funds. This implies that the initial estimate was incorrect and will certainly be an embarrassment to the design team. The presentation to the client of a carefully completed and detailed budget is the most convincing argument available to justify the necessity for additional funds.

Question 4.3: Arguments For and Against the Briefing Stage as the Most Important Stage in the Development Process

The whole thrust of this chapter and the previous one has been to emphasise that the briefing stage is one of the most crucial stages in the life of a building project. In particular, it is the time when most of the important decisions that influence up to 70–80% of the total costs are taken. Also in terms of the time and cost expended on the project, whilst the briefing stage probably represents only 3–8% of the time and costs, the major part of the project costs will be largely committed by the essential decisions taken during this stage.

If you still have doubts and concerns about this proposition then go Ferry and Brandon (1997), Gray and Hughes (2001). These are not the only authors that make this point, and if you still have any lingering doubts then consult the further reading noted at the end of the chapter.

Therefore, I would hope that you agree with the statement in the question and the view taken throughout the chapters so far.

When you frame your argument of support it should revolve around the type of decisions taken during this stage, and specifically the ones which were discussed in Chapter 3: total height, storey height, circulation areas, plan shape, size of building(s), cellular or open plans, mechanisation/ prefabrication and redundant performance.

However, there may be times when the actual activities of the briefing stage become detached from the usual timing put forward in these chapters and in the texts. This may occur due to delays that create the need to condense time and the activities. Whilst the briefing stage should occur during the early part of a project as noted earlier, the actual activity may have to take place during the later stages because of various delays due to financing problems, difficulties with site acquisition, planning problems, late appointment of consultants/advisers, and the like. Do not confuse a later than expected briefing stage in these situations with the later stages themselves. The important decisions will still take place during the briefing stage irrespective of the extended nature of the design process, or the need to condense or overlap the stages.

Examples
If you are working in an architect's or quantity surveyor's office where you are involved in these activities, I would expect that you could review one or more projects and trace the development of design decisions through documentation and drawings on the pre-contract project file. This should be an interesting exercise to compare practice with theory, and if they do not match up then find out why.

If you are without such ready access to this information, I suggest you make contact with a quantity surveyor or architect and conduct an interview with someone who is regularly involved in these types of activities.

If you are engaged in projects, it should be useful to start analysing your work and activities and to begin understanding the process a little better so that you can improve your performance. If you are not involved in these activities, it is necessary for you to find out what they entail so that you will understand them a little better for the time when you may become actively involved in them.

Question 4.4: Difference Between an Estimate and a Cost Plan

The differences are:

- A traditional estimate is a prediction of the cost of a design, and is specific for a particular design. The cost plan is a flexible overall allowance permitting changes in the design without needing constant revision.

- The cost plan is a statement of how the design team proposes to distribute the money available on elements of the building, and is formulated so as to postpone firm decisions on detail design and specification until a later stage. Thus, it is a flexible target allowing changes in the component costs that make up the total cost limit. With such flexibility it is *not* specific for a particular design.

- Once the estimate is prepared, it is seldom referred to until the time comes to compare it with the tender. The cost plan is referred to continually throughout the design process.

- In contrast, the cost plan is used continuously throughout the subsequent documentation (Detail Design) stage as a method of checking that the detail design is constrained within the cost limit. The cost plan is therefore a frame of reference. The cost plan helps the design team to detail the design within a cost framework.

Question 4.5: Low Quality Building and Cost Planning

The design team should not remain silent. In cases where remedial action would produce a building of low quality, the design team should ask the client group if they wish to release additional funds. It must be remembered that it is one of the aims of cost planning is to ensure a balance of expenditure between elements. This also applies to the project as a whole and if the design team consider that the quality of the project is being compromised then they must place the facts before the client.

In this undesirable situation, the cost plan is a powerful tool with which to argue the case for more funds.

Review Question

What are the three principles of cost control?
We hope you can answer this without any difficulty by now. Just in case you have not got the message, they are:

- There must be a frame of reference (cost limit and cost targets);
- There must be a method of checking (cost checking);
- There must be a means of remedial action (adjustments to cost targets).

If you are still having difficulty with these principles you must read this chapter again and also refer to the *further reading* given at the end of the chapter.

What are the main stages of the RIBA Plan of Work described in this chapter?
+1 Establishing the Need (Pre-stage A)

1. Options Appraisal (Stage A)

2. Strategic Briefing (Stage B)

3. Outline Proposals (Stage C)

4. Detailed Proposals (Stage D)

5. Final Proposals (Stage E)

6. Production Information (Stage F)

    ... and to finish off the pre-construction stages:

7. Tender Documentation (Stage G)

8. Tender Action (Stage H).

We will use these terms throughout this text, but you should be aware that other terms and expressions are used within the industry and the various professions that are different to those noted here. For instance, the briefing stage is sometimes referred to as the *Schematic Stage* and *Outline Sketch Plans*. The Detailed Proposals and Final Proposals can be referred to as *Sketch Plan Stage Scheme Design, Final Sketch Plans* and *Design Development.* The final stage in our series, the Production Information, is known as *Tender Documentation Stage, Detail Design, Working Drawings* and *Contract Documentation*. Whilst every effort should be made by all design team members to use the terminology used by the RIBA, in practice there may be deviations from this by other disciplines. The important thing is that you understand the stage they are referring to, and you can categorise it into the RIBA Plan of Work.

What are the six stages (plus one) required by the RIBA and briefly describe the cost planning aim of each stage?

• Pre-stage A	Establishing the Need
	Emphasis is on functional requirements and nature of client's problem. Is a building the solution? No detailed cost studies needed. Establishing the budget.
A. Options Appraisal	Identification of client's requirements and possible constraints on development. Functional, technical and cost studies prepared to enable client decide whether to proceed. Identify the cost of the preferred solution.

B. Strategic Briefing

Preparation of strategic brief confirming key requirements and constraints. This is a first cost indication to the client based on an outline statement of the client's needs in the strategic brief. Determine the target cost. The target or indicative cost is intended only as a guide for feasibility and planning purposes. The aim is to establish an initial budget for the client, or if already prepared, to confirm or reject the feasibility of the budget.

Value for money framework is established. Availability of finance may also be investigated.

C. Outline Proposal Cost

Commence development of strategic brief into full project brief. Prepare Outline Proposals and estimate of cost. Prepare initial cost plan. This is the first cost plan and estimate and is usually prepared on the basis of a client's brief, investigated site conditions and preliminary sketches. It may be necessary for the cost planner to evaluate the comparative cost of various outline proposals. The aim is to isolate the optimum design solution to satisfy the requirements of the brief.

D. Detailed Proposals

Complete development of the Project Brief. Preparation of Detailed Proposals. Application for full Development Control approval. Prepare firm cost plan. The first detailed elemental cost plan is prepared. The cost plan is based on the first detailed proposals and this confirms the cost limit.

E. Final Proposals

This stage prepares the final proposals for the project to permit the co-ordination of all component parts and elements of the project. Carry out cost checks of the design as it evolves against the cost plan. It represents a confirmation of the cost limit for the project. The aim is to ensure that the overall design is the most cost effective in satisfying the requirements of the project brief; to confirm or set the final budget; and to check the elemental cost targets.

F.   Production Information   Production Information is prepared in sufficient detail to enable tenders to be obtained. All statutory approvals must also be complete. This is the assessment of the lowest acceptable tender price for the work based on the completed contract documents. It should reflect all the factors currently influencing the level of pricing on the project. The aim is to ensure that the completed design is contained within the cost limit and the forecast tender sum by carrying out the final cost checks of design against cost plan.

Some of these descriptions are taken from the RIBA (2000) Plan of Work and you must refer to this document to obtain an authoritative statement of the activities and aims of each stage.

Chapter 5

Sources of Building Cost Data

It may be said – with about as much exaggeration as may be normally attributed to such statements, but no more – that the publication, 50 years ago, of the Ministry of Education Building Bulletin No. 4 introduced the concept of elemental cost planning to the UK construction industry.

Joe Martin, Executive Director,
Building Cost Information Service (2002)

Chapter preview

A pre-requisite of any kind of cost management system, including cost planning, is the need for reliable, cost data in the form of cost information which reflects the range of cost management being undertaken. In fact it was through the recognition of this need and that an abundant and readily available source of cost data existed, in a form which could readily be processed to form useful, extensive and reliable cost information that provided the catalyst for the cost planning process to be established as highlighted in the above quotation. It is worth going back a little in time to explore where this cost data source came from and how it became so useful in the cost planning process.

The development of cost planning, as an important and significant development of the expertise of the quantity surveyor, began in earnest during the 1950s as a result of the rapid expansion of construction activity within the UK, especially in education, after the World War II. Unfortunately the British quantity surveying profession, working as part of the design team, were often failing, sometimes with catastrophic consequences, to predict the likely cost of future buildings. As a result time wasting and unsatisfactory cost saving measures had to be put in place to bring about the necessary savings to ensure the project remained within budget. Unfortunately these cost cutting measures invariably led to a less than satisfactory outcome, as such measures were usually the substitution of cheaper and less satisfactory materials such as floor, wall and ceiling finishes. The finishing elements suffered the most because such changes were easier to make, as they did not impact significantly on the basic design, as would be the case if changes in shape, content or uses of space were made. Alterations of this nature not only meant considerable additional work for the design team, especially the architect, but also the successful contractor whose selected method of carrying out the work may have to be reviewed and changed.

So why was it that the budget estimates were often incorrect? The answer probably lies somewhere in a combination of the following features that were prevalent at that time:

- Inflation was becoming a more significant feature of the British economy, which of course meant that cost prediction into the future was becoming more difficult to carry out with reasonable accuracy.

- There was a rapid increase in construction activity that led to mismatches in supply and demand. Thus, the supply of contractors to meet the increase in demand for new construction work could not be met, which led to increased tender bids by the tendering contractors that again led to less objectivity and accuracy in cost prediction.

- A further pressure on cost prediction was the fact that construction solutions and choice of materials became more numerous and varied. This again made cost prediction more difficult as there was less familiarity in terms of design and construction solution availability.

Before looking at how the quantity surveying profession reacted to this challenge it is worthwhile, in order to put this chapter into context, to remind us of the tendering environment within the UK, at that time.

The majority of building contracts, especially those let by central and local governments, were lump sum firm price contracts, based on a standard form of contract, together with architect's working drawings and bills of quantities. Usually such contracts were let in open competition, that is, as many tenderers who wished to submit bids could tender.

Today, such contracts are being used less and less (RICS, 2002) due to the fact that they have tended, often, to lead to breakdowns in relations between the designers, including the client, the contractor and subcontractors. Also sub-optimal designs are often selected as the buildability of the design solution in terms of its construction methods and materials are ignored, or not fully understood.

So we have now identified the reasons for poor accuracy in tender bid prediction and consequently the reasons why the estimating methods that were deployed at that time were no longer working successfully. It is worth mentioning that the methods of estimating being deployed had no measure of control to ensure, before tender submission, that the estimate was still valid. Such estimates were based on a variety of techniques such as superficial and cube methods, functional unit methods and approximate estimating (these will be discussed in a later chapter). There was a further drawback in the use of these techniques, in that value for money was not always being obtained as the correct and balanced distribution of costs, a feature of current cost planning techniques, was invariably not achieved.

Response to Problems

As a result of these failures, it was necessary to devise a more successful method of predicting costs and ensuring that value for money was achieved. Initially a reactive approach was taken, in order to account for what a building was costing. This approach arose because central government, which was providing funding for buildings including schools and housing, wished to know why such projects, ostensibly of a similar nature, were costing substantially more or less in one local government area compared with another. As a result, the notion of elemental cost analysis was developed as a means of comparing one project with another, in terms of parameters of specification, geometry and size. The basis of such cost analyses was the notion that an element of a building would allow meaningful comparisons to be made between different parts of buildings in a consistent, objective and comprehensible way. As a result, the idea of functional elements was developed, which fulfilled this objective. (Elements were defined in the previous chapter.)

More detail will be given later in this chapter about the nature, purpose and form of elemental cost analyses but as an example, a typical functional element is that of roof which as can clearly be seen, provides the same function in a building independent of quantity, configuration or specification and thus allows meaningful comparisons to be made. Thus, armed with this cost information, derived from the bills of quantities, some outline specifications and outline graphical information, it could be easily ascertained as to why the roof in one building costs more or less than that in another. Some reasons may include the following:

- For example: is one roof bigger than the other? Is one more complex than the other? Pitched, flat and subdivided into different configurations?

- Is one roof a different specification, for example concrete, steel, timber, slate, concrete tile or felt coverings?

- At what time were the bills priced? Depending on the market conditions prevailing at a given time then naturally the prices submitted by the contractor will differ.

Of course the sources of these cost analyses were the priced bills of quantities obtained from the various lump sum tender submissions prevalent at that time. As such, these documents were an accurate reflection of the successful contractor's view of the cost implications of the particular project being considered, expressed in a logical consistent manner, as reflected in the quantities of work and the unit rates expressed against them. It is also worth remembering, that the priced bill of quantities themselves had been prepared in accordance with a nationally agreed set of the rules – the Standard Method of Measurement for Building Works. This standard approach to measurement was first introduced in 1922 and is now currently in its 7th edition (amended 2000), thus further increasing the reliability of the costs and data.

A slight complication was that it was necessary to reschedule the information contained in the bills of quantities as, generally, they were presented in trade or work section format, which was the most convenient arrangement in terms of pricing for the tendering contractors. The reason for this was that the bills contained all the similar pricing information, conveniently collected together. For example, the bill grouped together all the concrete work, all the brickwork, all the timber for joinery and carpentry, and so on. However, in terms of elemental cost analysis and deriving costs associated with each functional element, such costs are, in the main, independent of the type of work section or trade contained in the bill. That is, the items in a trade or work section do not coincide with the work required in an element. The items that make up a complete element may be distributed through a number of trades. For example, concrete work might be found in a variety of functional elements such as foundations, external walls, internal walls, floor slabs, roof slabs, etc.

So, getting back to the main story as to how cost planning developed as an important discipline performed by quantity surveyors and the pivotal role elemental cost analysis played in its establishment, it was now possible for an objective approach to be taken when analysing why one school building was perhaps costing more or less than another. Quite simply this was achieved by producing elemental cost analyses from the priced bills of quantities. This information could then be used to explain any differences in costs between one school project and another. As a result, when central government funding was being allocated to local authorities, it enabled the managers of this process to have more confidence that value for money was being obtained, which had not always been the case before.

It was quickly realised by a number of far-sighted quantity surveyors, that these elemental cost analyses, as used in the above important, but essentially reactive process, could also be applied in a much more dynamic modelling role. That is, to help predict and manage the future costs of buildings. You may recall that the need for better cost management of building projects came about due to the lack of ability to predict their future costs. The simple answer was to apply some lateral thinking to the problem and make the connection that an elemental cost analysis, a document reflecting factual information as derived from the relevant priced bill of quantities, could be adapted to reflect a normative model of the likely costs of a future similar building project.

This simple logical step now forms the basis of all design-based cost planning as carried out by the global quantity surveying profession, in varying degrees, taught in various building economics degree courses throughout the world and explained and demonstrated in various textbooks such as this one.

So how did it work? This will be explained and demonstrated in detail in this book, but in order to explain the development and use of elemental cost analysis, a brief summary is included here, as way of explanation of the process. The first step is to select a cost analysis of a building, which matches the building to be built, for example, a school as close as possible, in terms of specification, size and geometry. This is often called the process of *interpolation*. Once the nearest match possible has been obtained, the analysis can then be manipulated to produce a cost plan that states how much should be spent on each element. The manipulations performed are generally threefold:

- Adjustment for *quantity*: area of external walls in analysis, adjusted to match areas in the proposed building for inclusion in the plan.

- Adjustment for *quality*, in terms of different specifications between the analysis and the plan, that is facing bricks type A in the analysis, facing bricks type B in the plan.

- The final adjustment is for *time* (price). The analysis was carried out on a past tendered project and the future building is to be built at some time in the future. So, the cost information in the analysis is adjusted through the use of time-based indices.

In these early days little information was available, but now through the use of the Building Cost Information Service (BCIS) and other generally available sources of information, these adjustments can be performed with more confidence and reliability than during the earlier pioneering times. It is also worth reminding ourselves that, what is described above as being a straight forward process, is in fact, extremely difficult to carry out with a reasonable level of accuracy and needs considerable judgement and expertise by the quantity surveyor as well as the cooperation and integrated activities by the other members of the design team.

To summarise, initially cost analyses were used in order to provide objective information as to a building's costs, that is, a diagnostic tool. The potential of these analyses was then seen in cost management terms as a predictive tool by their manipulation to provide a cost plan; a statement of how much should be spent on a project (budget) and its various functional elements (cost targets).

Data Collection

The driver behind these developments was the failure of the design team and, specifically the quantity surveyor, to provide realistic cost estimates at the early stages of the design process within a lump sum competitive tendering environment. As a result of these developments it was felt worthwhile to identify a nationally agreed approach and to promote good practice within this emerging discipline in the UK quantity surveying fraternity. As a result a, 'think tank' was set up in the 1950s consisting of the leading exponents of the emerging discipline and they promoted two important directions for future developments:

- That cost planning based on the use of elemental cost analyses derived from priced bills of quantities be given formal backing and that the discipline of cost planning be an integral feature of the design cost control process within lump sum tendering. The data used in the production of these cost analyses was derived from the priced bills of quantities used in the lump sum tendering process.

- Support was given to the establishment of a nationally agreed approach to the preparation and presentation of elemental cost analyses.

As a consequence of these recommendations the cost planning discipline emerged in the UK and eventually became established. Cost planning is made up of the following stages reinforcing the basic principles enunciated earlier in Chapter 4:

- Setting a realistic *budget* (or cost limit) at the early stages of the design process (before the development of cost planning, this was the only stage carried out).

- Seeking value for money by considering various possible design solutions and setting *cost targets* for each element making up the proposed building.

- Providing a means of financial control by allowing the cost targets to be *cost checked* as more detail about the building is known, and ensuring *remedial action* is taken should there be a variation between the cost targets and the cost checks.

Each of these stages was recognised and described within the RIBA Plan of Work (RIBA, 1998, 2000) indicating how well established the cost planning process had become.

As a result of these recommendations and the obvious need for this approach from the industry, the Royal Institution of Chartered Surveyors (RICS) established the BCIS (RICS, 1969) on behalf of the building industry and its professions in 1961. The BCIS acted as a repository or data bank for elemental cost analyses, prepared by quantity surveyors working in private practice and local and central government, and allowed for their access by design organisations to help them in their cost planning work. The BCIS during its formative years was only available in hardcopy, relied on the goodwill and the public-spirited transparency of those organisations willing to submit their analysed bills. Access to the information was provided by means of an annual subscription, the rate of which varied depending on the size and nature of the organisation.

A crucial role of the BCIS at this time was to produce, on behalf of the industry, a Standard Form of Cost Analysis (SFCA) (RICS, 1969) that basically set out the following important parameters for the collection and presentation of building cost information:

- Principles of analysis.

- Definition of terms.

- The range of common elements, definition and content of each one.

- Instructions on how to carry out the preparation of the cost analysis from the priced bills of quantities from the successful tenderer.

The benefit of the development of the SFCA was to resolve issues of inconsistency and misunderstanding, both in their preparation and in their use. Thus, the establishment of the BCIS fulfilled two important roles:

- Facilitated a standardised approach to the preparation of cost analyses.

- Established a comprehensive library of cost analyses, held nationally for the benefit of the building industry, administered and managed by the quantity surveying profession.

It is worth stressing that these far-sighted developments were set up during a time when computing was virtually non-existent and, when it began to be introduced during the 1960s, was expensive, unreliable and definitely non-user friendly with its main frame computer department bureaucracy. There is no doubt that at the time it was a perceptive approach during a period when information and its management tended to remain in the domain of 'in house' or confidential. This was before today's advances in global communications, including of course the *World Wide Web*, had not been developed.

Initially, as an aid to the production of the cost analyses, the bills of quantities began to be prepared in elemental format, thus saving the time consuming exercise of manually reprocessing the work section or trade bills into an elemental format. Unfortunately, the tendering contractors were not enamoured by this approach, due to the fragmentation of the same work or items throughout the various elements. The power of the computer began to be harnessed, and despite the early difficulties, highlighted above, computers began to be deployed with increasing success. Bills of quantities then began to be produced by means of the computer. Many different formats of the bill could be produced with great speed and accuracy, including elemental arrangements, precluding the need to manually reformat the bill of quantities [Local Authority Management Services and Computer Committee (LAMSAC), 1970].

The developing science of cost planning, underpinned by the BCIS, gathered further momentum in the UK, as the new degree courses in construction economics and quantity surveying, invariably in their final year of study, included cost planning as a separate subject, with its own academic independence and integrity. By the mid-1970s cost planning was a well-established discipline in the UK and the BCIS was an integral feature of the cost management process. The BCIS continued as paper-based (using the standard mail service) for information transmission until in 1984 the BCIS introduced access to its database via a telephone modem link (BCIS, 1984).

This was a particularly exciting development, as it made the system much more easily and instantly accessible and dynamic. In addition to the on-line facility, a modelling capability was introduced as an aid to the cost planning process.

The modelling facilitated statistical analysis of data to provide more confidence in the cost advice being produced. An approximate estimating package was also introduced, which allowed the cost planner to build up their own computer generated cost plan from the analysis or analyses selected as a basis for the cost plan.

In 1998 the service was made more user friendly and compatible with current trends in information access by becoming web based, facilitating access to information contained on the BCIS website (*BCIS Online*) as well as allowing users to submit their information over the web. It was decided to abandon the cost-modelling feature of the earlier system, as it was felt to be too crude and unsophisticated to aid the cost planner in the cost planning process. However, no doubt, in the future, such a strategy will be revisited, as, with ever increasing sophistication in information technology, modelling, including cost modelling, will become more reliable, robust and beneficial in providing more objective appraisals.

It is also worth pausing for a moment to emphasise that most offices keep their own cost information, including cost analyses, to support their cost planning services. Clearly, such organisations would prefer, in an ideal world, to use their own cost information, as more trust can be placed in it since it is more familiar to them, and they can be more confident about its integrity and reliability. Unfortunately, even the largest offices will have difficulty in keeping and maintaining a comprehensive database of the range of building projects they are likely to be involved in and thus they are likely to have insufficient cost data to carry out effective cost planning. Even the *BCIS Online* does not and cannot contain cost analyses necessarily closely reflecting design and construction solutions for use in all future building designs. Of course, the more information that is available within its database, the more data it is possible to access and analyse during the cost planning process and the more useful the service becomes.

The Current Situation

Today the BCIS has developed into more than simply a collection of elemental cost analyses, prepared in accordance with a set of guidelines. It now forms a dynamic and integral feature of cost planning through its sources and data bank of capital cost information. It is now extensively used by all design and construction team members. This increase in its wider use has also been driven by recent and current trends towards the use of more integrated procurement strategies, incorporating more harmonious relationships between the various team members involved in the design and construction process. In fact, the BCIS has taken on board these changes to ensure that it is relevant to servicing the current needs of the building industry.

Currently the database comprises the following, all of which can be accessed by the users (*BCIS Online*).

Cost Analyses

Over 14,000 analyses are available across the complete range of building types. These can be searched against a comprehensive set of criteria to find the results, to match the user's project. The level of information available ranges from total building costs to full elemental costs and drawings.

Typical search criteria available are:

1. *Building Function*: Schools, libraries, clinics, etc. selected by means of the well-known CI/SfB system, by which buildings are classified within BCIS.

2. *Type of analysis*: Group elemental or detailed elemental.

3. *Type of work*: New build, refurbishment, extension.

4. *Number of floors*.

5. *Gross floor area*: A range of floor areas should be used that approximates to the proposed floor area of the project for which the cost plan

is being prepared. The chance of finding an exact match of floor area clearly, is unlikely.

6. *Location*: This allows the searcher to match the area in the UK of the analyses sought to the area in the UK of the building to be cost planned.

7. *Date*: The analyses sought are historic. Generally analyses will be sought, which are as recent as possible, as this will increase the reliability of the data, all other things being equal, for use in the cost planning process.

The objective of the exercise is to select an analysis or analyses that match as closely as possible the project which is to be cost planned, using appropriate search criteria, as outlined above. The search criteria can be refined or expanded depending on the success of the search being carried out. There are additional search criteria available, such as specification level, but the above criteria are generally used when carrying out this selection process.

How the analyses are selected and used in the cost planning process has already been explained and amplification is given elsewhere in the book. However, the reader is strongly advised to study extracts from the latest SFCA Principles, Instructions and Definitions (RICS, 1969, reprinted 2003) and the Detailed Form of Cost Analysis together with the BCIS Standard Elements (contained in these Appendices), which together fully explain the nature, form, purpose and application of cost analyses in the cost planning process. The complete version of these forms can be found on the website: http://www.bcis.co.uk

Tables 5.1–5.4 show a typical elemental cost analysis derived from the BCIS website. The analysis from the website is contained in one table, whilst for ease of understanding their content (the whole analysis) is divided into several parts making up the four tables.

Indices

It is essential that we have the means to adjust for differences in prices at different times. Building price indices allow us to carry out adjustments to building costs and market conditions from a base date to dates closer to the present. Cost data taken from a tender of a previous project is only satisfactory for the time at which the project was tendered and in that location.

The BCIS provides a full range of building cost, tender and output indices together with a range of derived regional tender price indices and trade price indices. The purpose of these regional tender price indices is so that

Table 5.1 Background Project Information: Primary School Extension

BCIS *Online* – Elemental analysis number XXXX		A – 2 – 0000
Building Function:	712 – Primary schools	
Type of Work:	Horizontal extension	
Gross Floor Area:	448 m²	
Job Title:	Extension to All Boys Junior School, Curdsand Way	
Location:	Matchbox, Yorkshire	
District:	Deleted	
Grid Reference:	Deleted	
Dates:	Receipt: 24 May 2001	
	Base: 14 May 2001	
	Acceptance: 26 June 2001	
	Possession: 9 July 2001	
Project Details:	Single storey, four-classroom junior school extension and remodelling of existing school. External works include precast block and macadam paths; play area and car park, stonewalls, metal fencing, landscaping, services, drainage and site lighting	
Site Conditions:	Gently sloping green field site with moderate ground conditions. Excavation above water table. Restricted working space and highly restricted access	
Market Conditions:	Project tender price index: 196 on 1985 BCIS Index Base	
Client:	Droppend County Council	
Tender Documentation:	Bills of quantities	
Selection of Contractor:	Selected competition	
Number of Tenders:	Issued: 4 Received: 3	
Contract:	JCT Local Authority Contract 1998 Edition	
Contract Period (Months):	Stipulated: 9 Offered: – Agreed: 9	
Cost Fluctuations:	Firm	
Tender Amended:	PC/Provisional sums revised	
Tender List:	£714,635	–	
	£782,915	+9.6%	
	£810,410	+13.4%	
	Contract breakdown		

(Continued)

Table 5.1 (*Continued*)

BCIS *Online* – Elemental analysis number XXXX			A – 2 – 0000
Measured Work	£429,017		
Provisional Sums	£16,400		
Prime Cost Sums	£183,600		
Preliminaries	£47,738		
Contingencies	£18,050		
Contract Sum	£694,805		

Source: Adapted from *BCIS Online*. Details of location and BCIS references suppressed for confidentiality reasons.

Table 5.2 Physical Project Description

Accommodation and design features: single storey four-classroom junior school extension to form courtyard. Trench fill foundations, PCC suspended ground slab. Steel frame. Forticrete facing block/block cavity walls; softwood cladding. Timber pitched roof, concrete tiles; steel flat and felt; patent glazing. Aluminium windows. Block, timber stud and glazed timber partitions. Flush doors. Plaster and tiles to walls; vinyl, lino, carpet and tile flooring; plasterboard and suspended ceilings. Fittings. Sanitary ware. Gas HW central heating, ventilation, electrics. Step lift. Fire fighting, alarms.			
Basement	$0\,m^2$	Area of external walls	$399\,m^2$
Ground Floor	$448\,m^2$	Wall to floor ratio	89.06%
Upper Floors	$692\,m^2$		Average storey heights
		Basement	m
Gross Floor Area	$448\,m^2$	Ground	m
		Upper	m
Usable Area	$348\,m^2$		
Circulation Area	$54\,m^2$	Internal cube	m^3
Ancillary Area	$28\,m^2$		
Internal Divisions	$18\,m^2$	Spaces not enclosed	m^2
Gross Floor Area	$448\,m^2$	Number of units	1
Functional Unit	Rate	Storeys as % of Gross floor area	
$348\,m^2$ Usable Floor Area	£1,598	Single storey	100.00%

PCC = precast concrete.
HW = hot water.

Table 5.3 Cost Analysis

Analysis					
Figures in *italics* rebased to a Tender Price Index of 225 (1st Quarter 2006 (Forecast: 225)) and adjusted using a location index of 110 (Berkshire (Index: 110, Sample: 133))					
Element	Total Cost	Cost per m²	Element Unit Quantity	Element Unit Rate	%
1 Substructure	*£84,387*	*£188.36*	m²		12
2A Frame (Includes Others)	*£12,228*	*£27.29*	m²		1
2B Upper Floors			m²		
2C Roof	*£71,635*	*£159.90*			10
2D Stairs					
2E External Walls	*£51,114*	*£114.09*			7
2F Windows and External Doors	*£53,920*	*£20.361*	m²		7
2G Internal Walls and Partitions	*£32,877*	*£73.39*			4
2H Internal Doors	*£15,319*	*£34.19*	No		2
2 Superstructure	*£237,093*	*£529.23*			34
3A Wall Finishes	*£14,708*	*£32.83*	m²		2
3B Floor Finishes	*£21,669*	*£48.37*	m²		3
3C Ceiling Finishes	*£11,465*	*£25.59*	m²		1
3 Internal Finishes	*£47,842*	*£106.79*			6
4 Fittings	*£16,452*	*£36.72*			2
5A Sanitary Appliances	*£6,417*	*£14.32*			
5B Services Equipment					
5C Disposal Installations	Included in element 5F				
5D Water Installations	Included in element 5F				
5E Heat Source					
5F Space Heating and Air Treatment (Includes Others)	*£57,000*	*£127.23*			8

(Continued)

Table 5.3 (*Continued*)

Element	Total Cost	Cost per m²	Element Unit Quantity	Element Unit Rate	%
5G Ventilating Systems	Included in element 5F				
5H Electrical Installations (Includes Others)	£47,250	£105.47			6
5I Gas Installations	Included in element 5F				
5J Lift and Conveyor Installations	£7,750	£17.30			1
5K Protective Installations	£763	£1.70			
5L Communications Installations	Included in element 5H				
5M Special Installations					
5N Builder's Work in Connection	£8,926	£19.92			1
5O Builder's Profit and Attendance	£3,000	£6.70			1
5 Services	£131,106	£292.65			18
Building Sub-total	£516,880	£1,153.75			74
6A Site Works	£53,639	£119.73			7
6B Drainage	£21,400	£47.77			3
6C External Services	£11,560	£25.80			1
6D Minor Building Works	£25,538	£57.00			3
6 External Works	£112,137	£250.31			16
7 Preliminaries	£47,738	£106.56			6
Total (Less Contingencies)	£676,755	£1,510.61			97
8 Contingencies	£18,050	£40.29			2
Contract Sum	£694,805	£1,550.90			100

Note: Percentages rounded off to nearest whole number.

Table 5.4 Project Specification

	Specification	
	Element	**Specification**
1	Substructure	Trench fill foundations. PCC beam and block floor with insulation.
2A	Frame	Steel columns and beams 3.52 tonnes.
2C	Roof	Mainly Redland Mini Stonewold tiles on timber pitched roof. Part stainless steel clad and felt flat roof. Patent glazing. Glue-laminated beams. Timber fascias and soffits. Aluminium rainwater goods.
2E	External Walls	Forticrete Bathstone block outer skin with insulated cavity and block inner skin. Steel lintels. Movement joints. Some timber cladding.
2F	Windows and External Doors	PC sum £45,000 for aluminium windows and doors.
2G	Internal Walls and Partitions	Block, timber stud and glazed timber partitions. 4 No. WC cubicles.
2H	Internal Doors	Timber doorsets and glazed screens with ironmongery.
3A	Wall Finishes	Plaster and emulsion; ceramic tile splashbacks.
3B	Floor Finishes	75 mm screed. Vinyl sheet, lino, carpet, clean-off matting and ceramic tiles. Painted timber skirtings.
3C	Ceiling Finishes	Emulsion to plasterboard and skim. Suspended ceilings.
4	Fittings	PC £10,000 for furniture supply; pinboards, cloak rails and sundry shelving.
5A	Sanitary Appliances	17 No. sanitary appliances.
5B	Services Equipment	PVC soil and waste pipes.
5D	Water Installations	Mains water.
5F	Space Heating and Air Treatment	Gas fired hot water central heating with radiators.
5G	Ventilating Systems	Mechanical ventilation to toilets.
5H	Electrical Installations	Electric light and power.
5I	Gas Installations	Gas supply to boiler.
5J	Lift and Conveyor Installations	Step lift.
5K	Protective Installations	Fire extinguishers.

(Continued)

Table 5.4 (*Continued*)

	Element	Specification
5M	Special Installations	Fire and intruder alarms.
5N	Builder's Work in Connection	Holes, painting pipes, etc.
5O	Builder's Profit and Attendance	2.34% of the services amount.
6A	Site Works	Demolition of toilet block. Site clearance. 510 m^2 macadam paths/play area and 68 m^2 car park; 10 m^2 block paving, 262 m concrete edgings/kerbs. 56 m natural stonewall. Site layout and seeding.
6B	Drainage	400 m of 100 mm clay ware drains, connection to existing. 5 No. brick and 3 No. PCC chambers.
6C	External Services	PC £5,000 for gas main. 182 m service trenches. Junction boxes. 10 No. lighting columns.
6D	Minor Building Works	Work to existing.
7	Preliminaries	7.59% of remainder of contract sum (excluding Contingencies).
8	Contingencies	2.87% of remainder of contract sum (excluding preliminaries).

PCC = precast concrete.
PC = Prime Cost.

adjustments can be made for different locations within the UK. For example it is more expensive to build in London than in the northeast of England. These indices allow such differences to be accurately taken account of. Similar indices apply also in Australia as well as other parts of the world where tender prices vary considerably in different regions due to a variety of factors such as climatic, economic, state of the infrastructure, etc. Other useful indicators are also available including the retail price index and output and new order figures. Again the nature, form, purpose and application of these various indices are explained later in the text. Included below is a summary of the BCIS cost and price indices.

Building cost indices
Nine indices from a 1985 mean base are updated monthly:

1. *BCIS Brick Construction Cost Index*

2. *BCIS Concrete Framed Construction Cost Index*

3. *BCIS General Building Cost (excluding M and E) Index*

4. *BCIS General Building Cost Index*

5. *BCIS Labour Cost Index*

6. *BCIS M and E Cost Index*

7. *BCIS Materials Cost Index*

8. *BCIS Plant Cost Index*

9. *BCIS Steel Framed Construction Cost Index.*

Tender price indices
Fourteen price indices from a 1985 mean base are updated quarterly:

1. *BCIS All-In Tender Price Index (TPI)*

2. *BCIS Firm Price TPI*

3. *BCIS Housing Refurbishment TPI*

4. *BCIS Housing TPI*

5. *BCIS Market Conditions Factor*

6. *BCIS Non-Housing Refurbishment TPI*

7. *BCIS Public Sector TPI*

8. *BCIS Refurbishment TPI*

9. *BCIS Rest of UK Sample TPI*

10. *BCIS Southern and Eastern Sample TPI*

11. *Fluctuating Price TPI*

12. *Private Commercial TPI*

13. *Private Industrial TPI*

14. *Private Sector TPI.*

The BCIS also has a number of indices that reflect costs (or inputs) rather than the price-based indices listed above. The indices reflecting the cost of different forms of construction and resource inputs are:

- general building cost
- steel framed
- concrete framed
- brick
- general building costs (excluding mechanical and electrical)

- mechanical and electrical
- labour
- materials
- plant.

The main point to note is that there are two basic categories: *cost indices* that measure input prices to the contractor (costs of labour, materials and plant) and *tender price indices* that measure the trend of contractor's pricing levels in accepted tenders (the cost to the client). It is also worth noting at this stage that these indices measure changes in costs and prices not actual costs and prices. As will be highlighted later in the book it is vital that the correct indices are used when making adjustments for costs and prices against time. Generally we will be using tender price indices when making adjustments as generally in our cost planning because we are dealing with the cost to the client (tender prices) and not input costs to the contractor. The graphs in Figure 5.1 taken from the BCIS website illustrates why the correct index must be used.

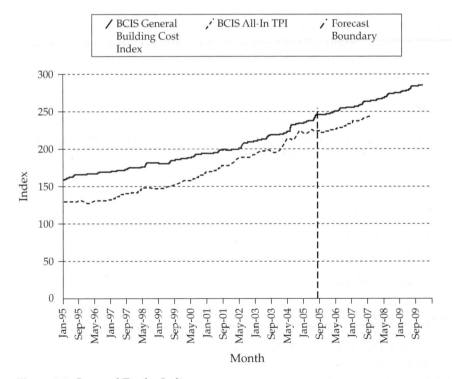

Figure 5.1 Cost and Tender Indices
Source: *BCIS Online* website (http://service.bcis.co.uk).

Building costs can also vary between locations due to one or more of the following factors:

- materials/products available or commonly used;
- ground conditions normally encountered;
- material/product prices due to distance from place of manufacture or distribution;
- distance from labour source;
- local regulations;
- labour productivity (e.g. due to adverse climatic conditions);
- builder's risk and market conditions;
- labour rates – due to different awards, state government taxes or insurance rates.

It can be appreciated that in a single index figure is contained a wealth of complexity. The choice of appropriate indices should be approached with care and an understanding of what they were measuring at the time on the previous project and what is being measured now. This is an extremely difficult choice to make! Proceed with caution. A poorly chosen index value can undermine a significant amount of research and accuracy contained in the estimate leading to the index adjustment.

Average Prices

Average prices for approximately 550 building types can be examined in a variety of ways using the following parameters:

- $£/m^2$
- Group element $£/m^2$
- Element $£/m^2$
- Element unit rates $£/m^2$
- Functional unit prices $£/m^2$.

Table 5.5 contains extracts from the BCIS website showing $£/m^2$ for new build offices and in Table 5.6 are the functional unit prices for new offices. Such information, together with the other average price information mentioned above are particularly useful during the early stages of the cost planning process when financial budgets are being considered.

Table 5.5 Office Costs Per Square Metre

Cost per square metre study							
Rate per m^2 gross internal floor area for the building excluding external works and contingencies and with preliminaries apportioned by cost. Last updated 18 March 2006. At 4Q2005 prices (based on a Tender Price Index of 223) and UK mean location.							
	£/m^2 gross internal floor area						
Building Function	Mean	Lowest	Lower Quartile	Median	Upper Quartile	Highest	Sample
New Build 320 Offices							
Generally	1,207	353	899	1,122	1,401	3,504	795
Steel Framed	1,179	353	914	1,124	1,339	3,504	334
Concrete Framed	1,374	637	1,034	1,319	1,627	3,135	267
Brick Construction	1,017	518	790	926	1,180	2,604	168
Timber Framed	922	671	772	905	1,074	1,170	13
Air-conditioned	1,374	518	1,054	1,287	1,593	3,504	281
Not Air-conditioned	1,038	521	806	983	1,198	2,400	269
1–2 Storey	1,033	353	800	970	1,207	2,604	346
3–5 Storey	1,237	521	969	1,168	1,401	3,504	307
6+ Storey	1,657	660	1,308	1,584	1,998	3,135	122

Table 5.6 Functional Unit Rates for Offices

Functional Unit Prices							
Functional unit rate for the building excluding external works and contingencies and with preliminaries apportioned by cost. Last updated 1 April 2006. Adjusted to a location index of 107 (Hampshire (Index: 107, Sample: 289)).							
	£/functional unit						
Building Function	Mean	Lowest	Lower Quartile	Median	Upper Quartile	Highest	Sample
New Build							
320 Offices							
m^2 Usable Floor Area	1,480	616	1,132	1,376	1,683	3,687	166
m^2 Net Lettable Floor Area	1,860	678	1,300	1,781	2,135	3,665	2
Number of Persons	24,938	13,004	15,309	18,562	25,148	64,313	8

Briefing, News and Digests

BCIS Online includes a briefing on the construction economy together with a 5-year long-term forecast of costs and prices as well as the latest news on building cost related issues.

Studies

BCIS Online undertakes a number of studies to investigate the effect on tender prices of factors such as location, regional trends, selection of contractor, building function, building height, type of work, site conditions, etc.

Dayworks

BCIS Online provides standard hourly rates for all of the main types of operatives in the industry both historic and current. Also details of construction wage agreements, national insurance and Construction Industry Training Board (CITB) levy are provided.

BCIS Online therefore provides the construction industry with the following significant user applications:

- Cost planning including early cost advice and development appraisals.
- Comparisons between and benchmarking against similar projects.
- Adjusting costs to different times and locations.
- Up-to-date information on the latest construction indices, statistics and daywork rates, construction news.
- Insurance rebuilding costs and valuations.

It is important to stress that the *BCIS Online* and all its earlier versions relies on a subscription. In addition to the subscription, the system can only remain up-to-date and comprehensive through the users' submission of information, either on-line or by other means of communication, such as the postal system. Data can be submitted as analyses at elemental, group elemental or total building level or as unanalysed contract documents, including priced bills of quantities, contract sum analyses and target costs from a variety of procurement strategies currently being used in the UK.

Since the Latham report (1994) and more recently the Egan report (1998) there has been a move away from the traditional lump sum competitive tendering strategies, that used to dominate the UK building industry, towards more integrated, harmonious, procurement strategies, such as Design and Build and Prime Contracting, incorporating partnering and framework agreements (RICS, 2002).

The *BCIS Online* system is willing to receive, in addition to information from traditionally tendered building projects, the following information:

- Contract sum analyses from Design and Build projects. Usually they will provide the cost analysis and the employer's requirement and contractor's proposal, the specification and the building area.

- Negotiated contracts as they can still provide a valid form of cost analysis.

- Contracts based only on drawings and specifications may also provide a cost analysis. Depending on how the contract sum analysis documents are structured, the pricing document may provide the basis for a cost analysis together with the specification, including the preliminaries with additional and supporting information.

- Where partnering and framework agreements are used, the cost plan or target cost agreed by the parties are often in elemental form. These can form acceptable cost analyses, provided they are clearly identified as such

BCIS, in 1996 in recognition of changing procurement strategies published *BCIS Elements for Design and Build: A Guide* to the use of elements for structuring contract documentation on design and build projects (BCIS, 1996) for use with Design and Build procurement arrangements. The aim of this document was to encourage Design and Build contractors to allocate their cost information in accordance with the BCIS elements in order to provide cost information, for the benefit of the industry, by its inclusion within the BCIS database.

The *BCIS Online* have also produced an SFCA for civil engineering, which attempts to define a functional elemental breakdown, together with appropriate cost carrying units for each element, together with an overall functional cost, such as cost per metre of tunnel, and interestingly, cost per tonne of treated sewage by a treatment works, and other similar useful units (Jaggar, 1997). Again this was developed in order to help create a more effective cost planning approach in civil engineering by developing a standardised approach to analysing cost information contained in bills of quantities. This approach remains of limited use, probably because the number of quantity surveyors involved in civil engineering contract administration, including the preparation of bills of quantities, remains small.

Question 5.1
Sources of Cost Data

Discuss the varying nature and form of cost data needed through the cost planning process?

Check some building price books and see if they provide good coverage in terms of the nature and form of data needed in the cost planning process.

If you have access to the BCIS identify the sections that provide this data.

In addition to the *BCIS Online* system, the RICS also publish the Building Maintenance Information Service (BMIS) *Occupancy Cost Information Service*. This service provides subscribers with regular, unique information on current, historic and forecast costs of building maintenance and occupancy. This parallel service provides assistance in the following areas:

- budgeting
- planning
- forecasting
- estimating
- cost control
- measured term contracts (usually deployed on ongoing maintenance contracts)
- valuations of dilapidations (due diligence) and life cycle costing.

It is likely that this service will become more in demand in the future as a result of moves towards Private Finance Initiatives (PFI) procurement strategies and the growth in facilities management, where the use and cost of the built asset throughout its life becomes an ever-important consideration.

As emphasised above, the *BCIS Online* and all its predecessors were established from cost data contained in priced bills of quantities. There are also a number of national price books, which contain cost information for use in estimating, including cost planning. These have arisen from a few large quantity surveying practices initially producing 'in house' cost information for their own use and realising that it could form a publishable document for use by the building industry and its associated professions, suppliers and subcontractors. Thus, these publications became a marketing tool and a commercial arm to their activities as well. Perhaps most important to the particular organisation concerned was its role as a marketing tool. Today, in addition to the annual building price publication as a hard copy book, the data increasingly is now provided on an accompanying CD for computer use.

The structure and content of a typical price book, as for example in *Spons Architects' and Builders' Price Book* (Davis Langdon, 2006), now in its 131st edition, is as follows:

1. Part I: Fees and dayworks
   This section gives details of the various fees recommended to be paid to the various consultants and for the services they provide.

2. Part II: Rates and wages
   This gives details of how much various operatives in the building industry are paid, together with breakdowns of the various components of the labour rates.

3. Part III: Prices for measured work
   This section gives detailed breakdowns in terms of labour and material for carrying out building work. The section is structured in accordance with the work sections and rules of the Standard Method of Measurement (RICS, 1989). Similar details are given for the preliminaries, which reflect general overheads including supervision and plant. The section itself is also subdivided into major and minor work.

4. Part IV: Approximate estimating
   This section gives strategic cost information in order to help with early cost estimating, including cost planning. It contains, amongst other things, cost and tender indices, building prices per functional unit and per square metre, together with cost limits for various building types, as well as information on government taxes.

5. Part V: Tables and memoranda
   This is a miscellaneous section, which includes general useful information such as formulae, conversion tables, planning parameters, design loadings, etc. together with useful addresses of many key organisations in the building industry.

A CD is included with the document, which includes the information contained in the hard copy documentation.

The above is atypical of similar price books as, for example, produced by the annual publications of *Laxtons Building Price Book* in the UK (Johnson, 2006) and *Rawlinsons Australian Construction Handbook* in Australia (Rawlinsons, 2006).

Chapter Review

The *BCIS Online* system remains an integral feature of cost planning within the UK and has benefited enormously from the development of the ubiquitous Internet access through the World Wide Web, which has made it cheap and easy to use and, most importantly, can be accessed any time from anywhere in the world. Thus, it has potential for development of similar approaches in other parts of the world, where such data is available, or could easily be made available. Of crucial importance to its success is that a group of community minded professionals is prepared to make a solid investment of their time and expertise to its development. It remains an invaluable tool for many construction organisations domiciled outside the UK, who may wish to consider breaking into the UK market, as the online system can be accessed and its database interrogated. There is a strong possibility that an international system could be developed, facilitating exchange of cost information and building intelligence pertaining to many construction industries throughout the world.

It is also worth stressing that the RIBA, through its well-established Plan of Work, recognised the importance of the cost planning process. Inspection of the RIBA Plan of Work shows how the cost planning process complements and is integrated into the design process and the framework indicates when, in the design process, cost planning is carried out. The RIBA itself has had to come to terms with the changing trends in construction procurement and this is reflected in the current edition of the RIBA Plan of Work (RIBA, 2000) for a *Fully Designed Building Project*. This latest update recognises and acknowledges the changing approaches to construction procurement by incorporating a revised Plan of Work for *Contractor's Proposals* (design and build). The reader is advised to study the current RIBA Plan of Work, extracts of which are contained in this text, to fully appreciate the various roles of those involved in the design and construction process and the need for integration and coordination between all those involved.

A further development arising out of the recognition that architects need to be able to give more accurate and reliable cost advice during the early stages of design (since October 2005) is the introduction of the RIBA/RICS cost calculator which is available exclusively to architects. The purpose of this service, available from the BCIS by annual subscription to architectural practices, is to provide the following:

- accurate and realistic cost information at project level
- cost guidance in a professional manner
- independent, authoritative cost information
- justify and further underpin fee payments.

Review Question

Discuss any limitations you feel may prevail in the cost planning process.

Do you feel that current changes in construction procurement strategies towards more harmonious and integrated approaches will impede or further develop the cost planning process.

References

BCIS (1984) *BCIS Online Users Manual*, The Royal Institution of Chartered Surveyors, London, UK.

BCIS (1996) *BCIS Elements for Design and Build*, The Royal Institution of Chartered Surveyors, London, UK.

BCIS Online (1998) http://www.bcis.co.uk

Davis Langdon (editor) (2006) *Spon's Architect's and Builders' Price Book 2006*, E & F N Spon, London, UK.

Egan, J. (1998) *Rethinking Construction*, Report from Construction Task Force, Department of the Environment, Transport and the Regions, HMSO, London.

Jaggar, D. M. (1997) *Civil Engineering Cost Analysis (CECA)*, Building Cost Information Services Ltd, London, UK.

Johnson, V. B. (editor) (2006) *Laxtons Building Price Book*, Elsevier Press, Oxford.

Latham, M. (1994) *Constructing the Team, Joint Review of Procurement and Contractual Arrangements in the UK Construction Industry*, Department of the Environment, HMSO, London, UK.

Local Authority Management Services and Computer Committee (LAMSAC) (1970) *Library of Standard Descriptions for Bills of Quantities*, LAMSAC, London.

Rawlinsons (2006) *Rawlinsons Australian Construction Handbook*, Rawlhouse Publications, Perth, Western Australia. Annual Publication.

RICS (1969 reprinted 2003) *Standard Form of Cost Analysis,* The Royal Institution of Chartered Surveyors, London.

Royal Institution of Chartered Surveyors (RICS) (1989) *Standard Method of Measurement,* 7th edition, The Royal Institution of Chartered Surveyors, London, UK.

Royal Institution of Chartered Surveyors (RICS) (2002) *Contracts in Use: A Survey of Building Contracts in Use During 2001,* RICS (Construction Faculty), London, UK.

Royal Institution of British Architects (RIBA) (1998) *Handbook of Architectural Practice and Management,* 4th edition, RIBA Publications, London, UK.

Royal Institution of British Architects (RIBA) (2000) *The Architect's Plan of Work,* RIBA Publications, London, UK.

Further Reading
Ferry, D. J. and Brandon, P.S. (1999) *Cost Planning of Buildings,* 7th edition, Blackwell Science, Oxford.
    Chapter 2: The impact of information technology, pp. 9–25.
    Chapter 13: Cost data, pp. 149–175.

Flanagan, R. and Tate, B. (1997) *Cost Control in Building Design,* Blackwell Science, Oxford.
    Chapter 8: Essential constituents of elemental cost analysis, pp. 91–117.

Jaggar, D., Ross, A., Smith, J. and Love, P. (2002) *Building Design Cost Management,* Blackwell Publishing, Oxford.
    Chapter 1: The context – definitions, historical influences and the basic approach, pp. 1–16.
    Chapter 2: Design cost management – the cost planning infrastructure, pp. 17–28.

Morton, R. and Jaggar, D. (1995) *Design and Economics of Building,* E & F N Spon, London, UK.
    Chapter 13: Cost prediction – science or guesswork?, pp. 289–314.

SUGGESTED ANSWERS

Question 5.1

The varying nature and form of cost data needed through the cost planning process?
The cost planning process is very much concerned with a progression from
coarse information to finer, more accurate information. This is because at the
inception or commencement of a building project very little may be known
about it. This may seem a severe limitation, but in fact is helpful in leading
towards a better overall solution as at this stage the variety of possible
solutions remain completely open and extensive. Once choices begin to be
made the range of solutions available becomes narrower and more limited,
and may in fact be sub-optimal as a better solution may not have been
identified through lack of time or poor research. For example, the range of
possible solutions should include all the possibilities of new build, adapt
existing, multi-storey, single storey steel frame, concrete frame, etc., but a
need to decide and curtail the process may mean a limited range of
possibilities is considered through *satisficing* behaviour on the part of the
decision-makers. That is, a sub-optimal, but acceptable solution.

Against this fluid and dynamic background information and cost data is
needed in a form that matches the particular stage (see the RIBA, 2000)
in the realisation of the completed building project design. It is also worth
reiterating that nearly all of the cost data used by the building economist has
been derived from priced bills of quantities used in the tendering process and
presented in arrangements that reflect the state of the design and
construction process in that particular procurement method.

Before analysing the varying nature of the sources cost information it is
important to stress that where ever possible construction economists will
prefer to use their own cost data sources as collected and classified within
their own organisations. The reason for this is that more reliance and
confidence can be placed on such personal data sources. Clearly it will often
not be possible to access all such data sources as even with the largest
organisation not enough data (representing a comprehensive range of
building projects) will be available, especially where results are required to
be backed up by statistical evidence. In fact, an interesting paradox that can
be noted is where statistical evidence that would be most helpful it is
generally not available. For example, not many Sydney Opera Houses,
Millennium Domes or Scottish Houses of Parliament buildings are available.
Conversely data about single storey primary schools are relatively freely
available.

Where in-house data is not available other sources of information will need
to be accessed, a prime source being the BCIS system as described earlier in
this chapter. This provides a comprehensive source of cost information

analysed and presented in a consistent manner. Other sources of information are various price books again mentioned in this chapter, which are a useful source of cost information for use by the building economist. Of course, there are many other sources of cost information ranging from technical magazines through to government and research organisation reports, which can be made use of in the development of the design and construction of the building project.

At the early stages of the design process cost data pertaining to the state of the design's development will be used such as functional cost data in the form of £/functional unit or £/m^2 gross internal floor area which can be used to establish a likely cost limit or confirm what function or space can be provided for a given cost limit. Examples of the nature and form of such data from the BCIS can be found earlier in this chapter. As the design develops further into the Outline Proposals stage cost analyses ranging from total building cost level, identifying quantity, quality and price information at project level, group elemental analyses giving information at group element level through to elemental analyses detailing information at elemental level as shown in the example included in this chapter

As the design solution further develops into stages D (Detailed Proposals) and E (Final Proposals) approximate quantities may also be used to identify elemental cost targets as well as for carrying the cost checking process as a means of carrying out any remedial action should it be necessary. Such cost data will be derived from information contained in priced bills of quantities that have been compressed to reflect a number of originally separately priced items with the same quantitative factors, but with different specification parameters in one composite item. For example, substructure may include excavation, hardcore, blinding, concrete, reinforcement and surface treatments all in one composite item. Such costs will be derived from office records or price books.

Additionally where cost information is not available due to this normally being carried out by specialists such as mechanical and electrical quotations, establishing cost targets and ultimately their cost checking will be obtained from specialist designers who at this stage will be part of the design team.

Once the design has been completed, tenders obtained and where lump sum tendering has been deployed the priced bills will be analysed to provide the original sources of data described in this chapter.

Of course indices are also available either from published sources such as BCIS or price books and some large organisations have prepared their own to give them more confidence in their reliability. The range of indices available that are published by the BCIS are listed in this chapter. As highlighted earlier selection of the correct index is essential if significant errors in cost prediction are to be avoided. As the procurement strategy is determined and decisions are made about whether the contract is to allow for fluctuations (price rises and falls) or not, a more sensitive choice of index can be made depending whether fluctuations are allowed or not.

Check some building price books and see if they provide good coverage in terms of the nature and form of data needed in the cost planning process.

If you have access to the BCIS identify the sections that provide this data.
If you have access to the BCIS website note the variety of ways the cost data is stored and presented. Notice the dynamic nature of the system where some limited cost modelling can be taken account of in terms of statistical analysis and manipulation of cost and time.

Review Question

Discuss any limitations you feel may prevail in the cost planning process.
Design cost planning described in this text book and other books dealing with this topic and as practiced by construction economists is based upon techniques and data forms that were introduced and developed many years ago (as described in this chapter). Although the techniques have been refined and further developed, especially with the introduction of information technology the principles remain little changed (frame of reference, means of checking and remedial action).

A limitation of current cost planning techniques is that they reflect the finished product of construction rather than the process of construction. This is essentially because our methods of collecting, analysing and presenting costs are based on units of finished work techniques that were established before the industrial revolution in the 19th century in the UK. That is, our measures or units of measurement tend to be static rather than reflect the dynamic nature and interaction of design, construction methods and technology.

Additionally when we present our units of finished work in our bills of quantities we tend to ignore the critical factor of location. Thus, our contract documentation (including the architects drawings) reflect the finished product – the building and it therefore makes it difficult for the constructor to produce his or her model of the processes involved in constructing the project. Essentially this involves identifying the activities involved, their relationship to each other, the resources involved and of course their costs. That is, a network and resource programme clearly identifying the critical operations and activities.

If you consult any project management textbook you will find explanations of how these complex tasks are carried out.

A number of attempts to overcome these limitations have been attempted which have had varying degrees of success. For a good treatise of these limitations and steps towards their rectification consult Chapter 12 in Jaggar et al. (2002: 141–165).

Do you feel that current changes in construction procurement strategies towards more harmonious and integrated approaches will impede or further develop the cost planning process?
A difficulty that construction economists face is that there has been a decline in lump sum tendering (often termed traditional contracts) as a result of cost, time and quality pressures. More open and integrated procurement strategies such as Design and Build, Prime Tendering, Alliancing and Private Public Finance arrangements, together with the introduction of partnering, have been introduced. So, the prime source of cost data (priced bills of quantities) is becoming less available. As a result the BCIS system may find that its database declines rather than expands or becomes stagnant. Additionally in house cost databases will also suffer a similar fate unless a concerted effort is made to collect and analyse data from projects using these newer procurement methods.

The BCIS has attempted to counter this possibility by incorporating into its database other forms of cost information arising from other procurement routes. These approaches have already been alluded to earlier in this chapter and which you are reminded to review again.

On a more positive note there is a move to greater transparency, within the design and construction process. As a result of these changing procurement strategies, together with the exponential developments in information technology, the ability to manage information and cost data more effectively, becomes more feasible. Such developments should facilitate the rapid assembly of specific information frameworks and requirements at the time it is required for the purpose it is needed. Thus, current generic information management strategies will become less widely adopted as information technology needs gather greater pace.

Such developments should greatly aid the work of construction economists. Greater understanding of the resource and cost implications of design and construction should be achieved as more effective model building (including more integrated resource and cost modelling) becomes available.

However, for this to happen the construction industry needs to become committed to introducing changes into its practices with less emphasis on a dispute driven culture and the recognition that transparency is of greater benefit to all stakeholders than opacity and professional obfuscation. This sentiment was summed up very succinctly in the Egan report (Egan, 1998)

> The Task Force wishes to emphasise that they are not inviting the UK construction industry to look at what it already does and do it better.
>
> They are asking the industry and government to join with major clients to do it entirely differently.

For further consideration of this interesting area of debate you are encouraged to consult various reports such as Latham (1994) and Egan (1998) that consider and recommend future directions for the construction industry.

Part B

COST PLANNING THE DESIGN STAGES

Chapter 6

Cost Planning the Pre-design (Briefing) Stage: Budget Setting

Most clients think of it as the final word. Many managers maintain it is the key word. Engineers will always read the word twice, just to be safe. Architects react as though it were a four-letter word. Quantity surveyors (QS) add a contingency and accept it may be a five-letter word. Whatever else it might be however, 'cost' has become a fundamental part of today's construction industry.

Sidney Newton (1989)

Chapter preview

We will now describe the practice of cost planning at our six plus one RIBA (2000) Plan of Work design stages. These stages take us from the inception of a project where we establish the need for the project to the production information and tender stage. Before we embark on a detailed description of cost planning practice, it is well to remind ourselves of the stages involved in the whole pre-construction process. In Chapter 2, we divided the cost planning of the design stages into six plus one key stages following the RIBA model of the pre-construction stages:

- Pre-design (Briefing stage) – Pre-stage A and Work Stages A and B.
- Outline Proposals – Work Stage C.
- Detailed Proposals – Work Stage D.
- Documentation stage – Work Stages E and F.

These are summarised in Table 6.1.

The outline of the approach used in this chapter and the following three chapters is also broadly summarised in Table 6.1. This table will provide a guide to the stages discussed in Chapters 7, 8 and 9 and will be used to identify the specific stage being described. These stages are now considered in turn. The shaded portion on the table indicates the stage under consideration.

The descriptions of each stage will emphasise the theory of cost planning given in earlier chapters and will provide an example of cost planning during each of these stages. In this way the theory can be related to the practice.

Table 6.1 RIBA Outline Plan of Work 1998

Pre-design (briefing)			Design			
Inception or Feasibility			Pre-construction Period			
Pre-stage A	Work Stage A	Work Stage B	Work Stage C	Work Stage D	Work Stage E	Work Stage F
	1	2	3	4	5	6
Establishing the Need	Options Appraisal	Strategic Briefing	Outline Proposals	Detailed Proposals	Final Proposals	Production Information
Chapter 6			Chapter 7	Chapter 8	Chapter 9	
Briefing			Outline Proposals	Detailed Proposals	Documentation	

Chapters 1 to 5 provided the theoretical foundations for cost management and the role of cost planning in achieving these aims. These principles will now be elaborated in the next four chapters. These chapters will concentrate on the practical aspects of cost planning. You may wish to refer back to the earlier chapters if you want to confirm the basis of an activity described in these applications chapters.

In simple terms the cost planning process can also be divided into the following sequential stages:

- set the budget;
- prepare the outline cost plan;
- prepare the detailed cost plan;
- cost checking and remedial action;
- tender documentation;
- analysis of tender submission.

As we progress through the various RIBA Plan of Work stages we are, in fact, following the simple model or framework defined above.

Whilst the stages outlined here are shown in a linear sequence, it should be remembered that in practice the sequence tends to be more fluid and iterative to reflect new or unexpected information becoming available to the design team. In these cases, the whole basis of decision-making up to that point may be thrown into question and the process may have to recommence or iterate to an earlier stage. Design teams will guard against these circumstances arising, but in some cases, the unforeseen may create some problems, or opportunities, that cause the design team to revise its decisions.

This chapter starts the cost planning process with the pre-design stage, which we have termed the briefing stage for simplicity and this stage consists of three Plan of Work stages:

- Pre-stage A – Establishing the Need.

- Work Stage A – Options Appraisal.

- Work Stage B – Strategic Briefing.

It is worth emphasising the importance of the Briefing stage. Many authors such as the Ministry of Public Buildings and Works (1964), United States General Accounting Office (1978), Latham (1994) Ferry et al. (1999), Jaggar et al. (2002) over an extended period of time have stressed that it is at the early stages of project where many important decisions are made that affect the economy, efficiency and timing of the project. The Construction Industry Development Agency (CIDA, 1993: 13) support this proposition in a publication which concentrates on the *project initiation* phase of a project:

> Rushing the concept and evaluation phase misses the best and lowest cost opportunity to get the project right. Changes made at the concept stage cost little but can have a major impact. In contrast, late changes are expensive or cannot be done, and disrupt and dislocate project delivery leading to cost and time overruns.

Therefore, it is vital that we use our best efforts to ensure that the client and the design team are served by the best information and decisions we can secure for this crucial early stage. However, at this stage, cost planning is at best, rudimentary and because of the lack of information, often non-existent.

Nature of the Problem

The production of buildings moves through a number of stages involving a large number of participants. The contributions from the formally constituted client, design and construction teams charged with the responsibility of delivering the project are crucial to the success of a project. However, it is increasingly recognised that external participants such as users, customers and members of the community may have a useful role to play in influencing the location, form, content and timing of the project. In fact, it is now appreciated that the multiple attributes that contribute to the success of a project (financial, social, technical, functional, aesthetic, environmental, political) are influenced by a whole host of decisions by various individuals, bodies and organisations (Property Services Agency, 1981; Barrett and Stanley, 1999). These early stages in particular can be categorised as a complex process (RIBA, 2000; Walker, 2002).

The decisions made during these early pre-design briefing stages in the life of a project are seen as a critical factor in influencing the fundamental characteristics of quality, cost and time of projects (Jaggar et al., 2002). Improving the performance of buildings in relation to these attributes should deliver better buildings that can play a major role in improving the effectiveness and efficiency of the economy in general (Murray and Langford, 2003). This is achieved through improvements in the quality and extent of goods and services provided through these better buildings now available to the community. Thus, governments have showed increasing concern and attention to improving the production capabilities and capacity of the construction industry. A proficient and efficient construction industry is seen as a key objective of economic policy in many countries. For example, government-sponsored agencies have been established in many countries to improve the performance of their country's construction industry. These authorities have stressed the importance of adequate briefing on projects and the need to invest more time and expertise on the early briefing stages of project development.

A decision to proceed with a building project is one of the significant decisions that should be reviewed (Woodhead, 1999). Also, it is during these early stages where most of the critical decisions that affect the economy, efficiency, timing, functional content, appearance and most important of all, the real value of the project. Numerous authors offer support for this proposition, including Dell'Isola (1982), Bennett (1985), Bon (1989), Ferry et al. (1999). A quote from one of these authors exemplifies the basis of these assertions:

> ... the planning and design phase of a building project deserves more abundant resources and closer attention because in this phase we simultaneously encounter the greatest possibilities for influencing the total project cost and the lowest expenditures associated with the project.
>
> Bon (1989: 62)

Research in the UK has identified the need for clients and their advisers to be aware of the importance of what can be commonly termed, the *strategic level of decision-making* (Keel et al., 1994; Gray and Hughes, 2001; Jaggar et al., 2002). Atkin and Flanagan (1995) conducted a survey of construction clients that indicated that the strategic level had the most potential for cost savings in a project. The importance of the strategic stages in the development of solutions was further reinforced by the Royal Institution of Chartered Surveyors (RICS) QS Think Tank (CSM, 1998a), which suggested that organisations require advisers who are willing:

> ... to challenge the *status quo*, who can tell them they have a problem because they have the wrong team, who will question what the project is costing and get better value for money.

The RICS, QS Think Tank (CSM, 1998a) also stated that client advisers should place high priority on:

- Understanding the project priorities and business objectives.

- Providing advice which assists clients to gain competitive advantage.

- Being client orientated rather than being focused too much on the details of the project to the detriment of the broader issues and objectives.

They considered that a growth area in the future would be project strategy work, although they conceded that few design professionals have yet established a strong position in this area (CSM, 1998b). With the advent of the RIBA Outline Plan of Work in 1998 (RIBA, 1998), this early stage was given due recognition by the definition of three new stages: Establishing the Need, Options Appraisal and Strategic Briefing.

Following this change, practitioners should now be devoting more time and apply more rigour to the decision-making during the formative stages of a project, particularly these pre-design stages. However, in practice, some obstacles have to be faced. The culture within the construction industry and of many clients often creates a climate of expediency. The consequence of this is a tendency to rush the early stages of projects and to eliminate or diminish the benefits of more care in the initiation or inception stage. Unfortunately, this short-term thinking does not always lead to a soundly based strategy to guide a well thought out project solution.

The Strategic stages of a project ideally should contain the following features (Friend and Hickling, 1987; Duckworth, 1993; Thompson and Strickland, 1995):

- Establish the need:
  - review and consider the need for organisational change(s);
  - consider the strategic environment of the organisation and its direction;
  - link strategic possibilities with organisational direction.

- Option development and appraisal:
  - identify strategic options;
  - decide on a strategic solution.

- Strategic briefing:
  - document the option(s);
  - client recommendation.

Once an organisation perceives a need for change in the way it conducts its present methods of business and operations then a process of making a choice about what that may entail takes place. This process may be explicit

and formalised by the relevant procedures in the organisation. If the
organisation is large enough it may have a capital works department (or
similar division) to carry out these specialised tasks. Alternatively, and more
likely in small- and medium-sized organisations, the decision may be implicit
and the organisation, individual or group may *muddle through* to arrive at a
decision to build (Woodhead, 2000; Woodhead and Smith, 2002). Whatever
the situation, once the crucial decision to build has been made then the
various possibilities should be explored to solve the client problem or to
satisfy the needs of the client. In addition, it must be recognised that most
clients involved in building development are likely to be occasional clients
who probably only ever build once and they tend to be small organisations,
not multi-layered national or multinational organisations (Masterman, 1992;
Kamara and Anumba, 2001).

In order to improve the process of project inception, the project team, which
includes the client, needs to look more closely at the nature of the problem
on which we have decided to focus. Rather than a rigid framework the client
and the project team requires a more general flexible schema to provide
the framework for establishing the *real* need of the organisation and
developing options. Using a problem-solving approach to guide this work
may be more useful. In general terms, a problem is a question in need of
solution. In the context of the organisation someone, or a group, has
identified an unsatisfactory state of affairs, which has created a need for
action. This need may be solved by the construction of a new building, the
extension to an existing building or a rearrangement of the activities and
facilities within an existing building, or some combination of these (Mohsini,
1996). Figure 6.1 summarises the strategic context in which this research is
being carried out.

Whilst everyone involved in the development and construction process
appears to agree that a sound and effective construction industry is essential,
it is therefore surprising that a lack of communication has been a persistent
theme in the literature for the last 50 years (Murray and Langford, 2003). The
UK provides a fertile field for relevant governmental and institutional
studies, analyses and reports. For example, reports have been published on a
regular basis in every decade from the 1940s beginning with the Simon
Report (HMSO, 1944), and most recently with the Latham Report (1994) and
Egan Report (1998). Similarly, in Australia, demands for a more efficient and
effective construction industry have paralleled similar concerns as the UK
(National Public Works Conference/National Building Construction Council
Joint Working Party, 1990 and DISR, 1999).

These reports have consistently focused on issues such as:

- Improving the organisation, management and coordination of
  members of the design and construction teams.

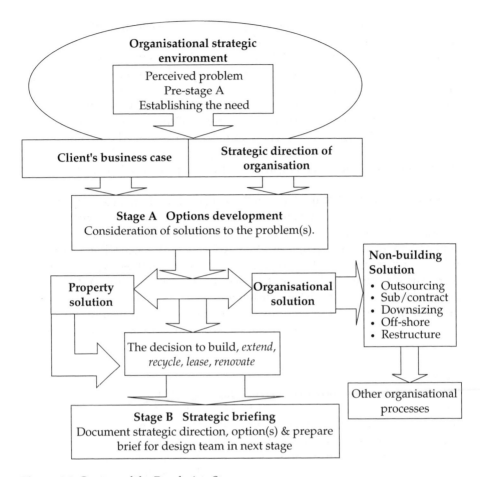

Figure 6.1 Context of the Pre-design Stage

- A lack of integration between architects, building professionals and the contractor.

- Poor communication between these teams and inadequate client consultation.

- The demand for more complex buildings from more exacting clients made the resolution of these issues more urgent.

The response of professions and industry has often been erratic, temporary and often too slow. Consequently, clients have demanded improved service and improved methods of procurement (Bresnen et al., 1990; Latham, 1994; Egan, 1998).

Evidence of this, in the UK, occurred when the largest grouping of private sector clients, the British Property Federation (BPF) seized the initiative in

1983, bypassed all the entrenched and self-interested construction professional bodies and introduced its own project organisation and procurement method (BPF, 1983). The BPF method emphasised the primacy of client needs and requirements with an independent project manager taking control of the design and construction process on behalf of the client. Government was similarly spurred to adopt alternative approaches to procurement (National Economic Development Office, NEDO, 1983; Department of Industry, 1982). These initiatives encouraged different and innovative project arrangements as well as procurement methods that are designed to align client needs with an improved project delivery process.

More recently in the UK the Latham Report (1994) and the Egan Report (1998) have provided a catalyst for the industry to become more client focused and to try to remove impediments to poor performance. In Australia, similar activities occurred (CIDA, 1993; AIPM/CIDA, 1995; DISR, 1999). Whilst conflict and disputes received most attention in these reports, special emphasis was given to the importance of clients, good briefing and the essential need for the industry to listen to the client(s). The strategic needs of clients did not receive any detailed attention, but the briefing process was identified as an important factor in the successful development of a project.

Hence, the clear message from countries such as the UK and Australia is that the construction industry is an essential sector of the economy. Its efficiency and effectiveness in providing buildings (capital works or assets) is crucial to the industries that rely upon it for the delivery of buildings, and to clients in the rest of the economy. The drive for greater productivity and competitiveness throughout these economies cannot overlook the construction industry (Morton, 2001). External forces throughout the economy are driving change and the quest for ever-greater efficiency. The industry, and particularly its professions, has been put on notice that they must improve their performance (Egan, 1998; DISR, 1999).

The supremacy of client needs in the process is now recognised by many of the practitioners in the industry. In fact, the development process that produces the final building is now subject to increasing review, analysis and change driven by the need to make it more responsive to the client, market forces and users.

Establishing the Need

After taking account of all the strategic factors and the organisational environment the decision to build represents a strategic initiative or assessment to embark on a built solution to satisfy the organisation's strategic objectives. It is a decision that members of the design team should be aware of, reflect on and consider why the decision to build was made. In this way, their advice and activities during the design stage should be more relevant to client needs. By understanding the history of the decision the design team may be better able to respond to client requirements and opportunities (Bilello, 1993; Bolman and Deal, 1997 and Woodhead, 1999, 2000).

However, most participants in the design process are often divorced from the decision to build within the client organisation and may lack the necessary background information that informed, guided or forced the choice of a building solution (Bilello, 1993).

Woodhead (1999) has conducted the most thorough and comprehensive analysis of the decision to build. He specifically attempted to explain a range of issues in relation to the decision to build: the process, the content, what influences it, why those influences affect the process and content. His research focused on the pre-design stage. That is, on the activities and processes prior to the client appointing the architect, design team or project manager. It identifies the typical steps leading to the (tentative) decision to build with the boundary separating the research from later design and procurement stages.

Strategic stimuli or triggers (Woodhead, 1999: 135) for the decision processes leading to a new building include:

- perceived opportunity or threat to the organisation;
- new ideas;
- changed location for operations or market;
- corporate finance issues;
- responses to other decisions.

Often the decision to build arises out of a need to *rescue* an implemented strategy. The trigger, or perceived opportunity, is a reaction to the recognition of a strategy that has previously failed, is failing, or is about to fail. The client considering the decision to build must often have to reconcile the issues related to organisational factors and be ready to adapt as it moves through the process.

After establishing the need for a project we move on to the next stage in the RIBA Plan of Work (2000). This is Stage A, the Options Appraisal stage.

Stage A Options Appraisal

Once the size and scope of the client's problem are understood during the pre-design stage, establishing the need, the client advisers should be in a good position to propose solutions, or options, to solve the organisation's problems and reinforce the client's business case.

A process is needed where participants who understand the client's problem and are aware of the organisation's needs can develop strategic options and then decide as to which of these options becomes the subject of documenting in a strategic brief. The conversion of the chosen option(s) into a working document for use by client, users, (stakeholders), business management and the design teams needs the use of a strategic document that can bridge the gap between the client, the users and the design team. This strategic brief with an emphasis on the required performance of the project should use a format, style and language all stakeholders and other groups can understand and be used as a base for the development of the project.

The process of creating options from an understanding of the problem in the context of the strategic environment, the client's business case and the strategic direction of the organisation is an important one that is decided by the client and the project leader at this early stage. There is no prescription for carrying out the work at this stage, and the client and the design team should agree on an approach that will create realistic options from their business case and then allow decisions to be made as to which option or options should be documented and developed in the next stage of strategic briefing.

Whatever process is decided for this stage it is likely to include a high content and interaction with the client team and possibly users as noted earlier. The command structure of the *expert* making a decision on behalf of the client is not likely to be appropriate, nor is it likely to elicit the best solution to the multi-faceted problem faced by many organisations. A workshop-based interactive approach at this stage is probably more suitable for this problem-solving and options generating context. As noted in Chapter 4 the range of approaches that can be used is many and varied.

Whatever approach is used it should suit the client environment and the problem context. All successful approaches are likely to be based on client or stakeholder involvement and clear communication of facts and decisions.

The RIBA (2000: 11) consider that the options appraisal stage:

> ... is simply to confirm that the objectives are reasonably attainable and affordable – in other cases the process may have to be more rigorous where factors – fundamental to the success of the project (need to be considered):
>
> - functional performance;
> - quality, architectural profile, corporate image;

- programme for operational use;
- available finance;
- value for money.

Once the client, users and the project team have agreed on a preferred option, or more than one option, the joint team should prepare a report comparing available and practical solutions, observations on a suitable procurement method and the range of professional skills required. This report can be the basis of client decision-making and the process can advance to the next stage, development of the strategic brief.

Stage B Strategic Briefing

The RIBA (1998: 184) defines the brief as a document rather than as a process of briefing as follows:

> The brief should be as comprehensive as possible, clearly expressed and accurate. It may be a formal document, used as a point of reference throughout the project, and as a yardstick by which the completed building is measured. No assumptions should be made. It is a professional duty to question the client's proposals and information and to identify ambiguities or omissions, establish a firm basis for progress, and reduce the risk of misunderstandings later. Brief-taking should not be hurried; any temptation to rush in with preconceived ideas and solutions should be firmly resisted.

The term *brief* is traditionally taken to mean the document describing the client's requirements for a building project, which is written prior to any formal design action; and briefing is the commonly accepted term for the process by which this document is produced (RIBA, 1998). However, the terms *brief* and *briefing* are also used customarily to refer to the ongoing process of eliciting and documenting the requirements of clients at various stages during the design of a building project (Kelly and Male, 1992). This view is also supported by Salisbury (1998: v).

> A brief is everything an architect needs to know about the building a client needs. The client's yearnings, ideas and vision should be clearly expressed in it, together with every activity and important piece of equipment or treasured possession to be accommodated – it is more than a verbal exchange of ideas. It is a creative act, which shapes the subsequent building, and it should be presented in the form of a well-constructed document, which is concise, realistic and as comprehensive as possible.

A considerable amount of activity in the UK, both practice based and pure research, has centred upon improving the briefing process. Much of this has been going on since the early 1980s and work by Newman et al. (1981), Goodacre et al. (1981, 1982a, 1982b), O'Reilly (1987), Cornick (1990), Murray et al. (1993), CIB (1997), Salisbury (1998), Gray and Hughes (2001) and Kamara and Anumba (2001), have firmly established client briefing as a process to be analysed and improved. Typically, the client and in some cases the architect are seen as the leaders of the design team and the *key* decision-makers who must have a process that supports their power and responsibility.

Gray et al. (1994) sees the initiation of the project as the most important phase. They see the briefing process in three stages:

- the statement of need;

- confirming the need;

- developing the functional or design brief.

The first stage they call the initial *statement of need*. This consists of a series of criteria or performance statements that form the basis for assessing and establishing the need for the building (Gray and Hughes, 2001: 93,94). The Statement of Need contains statements on important matters such as the function of the building, timing or programme for completion and an indication of priorities in the project. A design does not exist at this stage and the project is only considered in terms of its strategic approach, performance and function.

Through meetings and discussions the client confirms these statements. An important basic issue is to confirm the decision to build and that a building is the appropriate solution for the client's needs (Karma and Anumba, 2001). Therefore, this statement must be prepared in terms that the client can understand and can be readily transformed into a design concept for development by the design team. Following agreement of the strategic brief at this initial stage, eventually a *functional brief* can be prepared which is a comprehensive document that expands the performance brief and guides the design through the outline proposals and sketch plans stages (Gray et al., 1994 and Gray and Hughes, 2001). The functional brief document establishes the basic design policy and approach for the job. It precisely defines the design requirements (RIBA, 1998). There are many interpretations of the (functional) *brief*, which is the printed outcome of the client briefing process. The common thread that pervades most definitions is the prescriptive nature of their content. That is, it contains lists and descriptions of *what* will be incorporated in the design (Salisbury, 1998).

Overview of costing activities and documentation
The RIBA (2000) view this stage as being critical in terms of the development of the design. The client has a critical role in the strategic brief by specifying clearly all the key requirements and constraints.

The basis or information available at this stage consists of the brief and/or any sketches and other relevant information. The client is involved in confirming an extensive range of parameters and constraints that will form the basis of the outline design. These parameters involve the client and the project team and have been defined by the RIBA (2000: 14) as:

- preparation of the strategic brief;
- user requirements;
- schedules of accommodation;
- site information;
- design and material quality;
- facilities management;
- environmental services;
- sustainable development policy;
- whole life costing;
- timetable of critical events;
- target cost/cash flow constraints;
- procedures, time and cost controls;
- professional appointments;
- partnering;
- construction procurement;
- value management policy.

The cost planning result of the Strategic Briefing stage is a global cost figure that may be considered to be the first estimate. This first or indicative cost should be the current project cost plus escalation to tender date (if required) plus other costs as required, each separately identified.

The purpose of the cost planning at this stage is to document the strategic requirements of the project and plan future action. This preliminary cost advice is essential as its aim is to provide the client with a project appraisal and recommendation that assesses the project technically, functionally and financially. The cost advice based on the basic and preliminary information available at this stage is to establish an initial budget for client's organisation. The important action they have to take is to consider and confirm that the client's budget or cost limit is feasible.

The method of calculating this initial cost, based on very initial data is basic and can only be guided by the information produced by the client and other team members. So, at this early stage the project cost is built up using functional use(s) indicated. Where possible any attempt at assessing the gross floor area (GFA) of the proposal from the information provided in the strategic brief should be made. Functional unit costs from similar projects will be derived with external works and other identifiable costs shown separately.

It should also be noted that:

- Cost decisions made during the Strategic Brief stage are based on the client's functional requests, rather than on a particular design solution.

- Decisions made always take account of the limited nature of the information on which they are based.

- Decisions are always made in such a way as to impose minimal restriction on the range of possible design solutions later on.

The relationship between information available at the start of a stage and the design and cost decisions made during the stage is probably best described in diagrammatic form in Figure 6.2. As the available information (Box 1 below) expands and changes, it influences the interrelated aspects of design and cost decisions (Boxes 2 and 3). One aspect should not be considered without understanding the effect it has on the other two aspects.

In practice, if the cost planner is not involved until after the Strategic Briefing Stage A or the Outline Proposals stage B, then the two distinct stages A and B

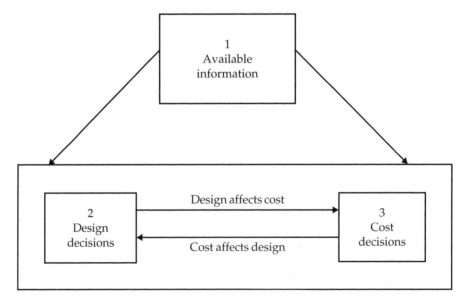

Figure 6.2 Relationship Between Information, Cost and Design Decisions
Source: Adapted from Ministry of Public Buildings and Works (1968: 179).

become one stage where the cost planner reports the Outline Proposal Cost. This is not good practice, nor is it recommended as the client may be receiving early cost advice (if any) at the budget setting stage that may be poorly researched, inaccurate and misleading. An unrealistic budget established in these circumstances is likely to create problems for the design team and the client for the remaining stages. This practice also goes against the emphasis of the importance of the Strategic Briefing stage described here and in earlier chapters, which contributes, greatly to the successful balancing of time, cost and quality in a project. It is a practice that may save design time, but could have far wider damaging cost implications for the total project.

The cost of a building is important to all clients. It does not matter whether the client is in the public or private sector, or whether the project is profit/income-generating or not for profit purposes. The estimation or calculation of the capital cost is a significant activity in the development process. The maintenance of the predicted cost through the various stages of the development process is of crucial importance to the client and lies at the heart of the cost control and cost planning process. The techniques of estimating employed must be reasonably accurate, appropriate for the stage of development of the project and be compatible with the quality and amount of information that the design team has produced.

Reflection
In Chapter 4, we emphasised the need to provide a frame of reference to cost planning activities. The estimate(s) of capital cost and the necessity to control the development of a design concept within the agreed capital cost estimate are crucial to the eventual success of cost planning and control.

The remaining part of this chapter now concentrates on defining and describing the various types and methods of preparing approximate estimates aimed at setting the budget of the capital cost, which will then be traced through the remaining design stages using a different set of techniques focused on distributing this budget. That is, we are examining the toolbox of techniques for cost planning. The suitability of these techniques through the various stages of a project is discussed in this chapter and the later ones of Chapters 7, 8 and 9.

THE TECHNIQUES: WHAT IS AN ESTIMATE?

Seeley (1996:154) captures the essence of the aim and purpose of design stage estimating as follows:

> The primary function of approximate estimating is to produce a forecast of the probable cost of a future project, before the building has been designed in detail and contract particulars prepared. In this way the building client is made aware of his likely financial commitments before extensive design work is undertaken.

Cost planning cannot succeed without competent cost estimating. Estimates are needed from the inception of a project, through all the design stages (previously identified in Chapters 3 and 4), up to the receipt of tenders, which should confirm the accuracy of the estimating on the project. Estimates may have to be prepared on the most meagre of information at the start of the project, or on detailed working drawings, specification and bills of quantities, just prior to the tendering stage.

Estimates may be prepared in minutes on the back of the ubiquitous envelope with little or no significant information, or may take days and weeks based on the most extensive and complex information. Clients and design teams will act on the advice derived from such estimates, which can result in changes of design, layout, content, materials, and in extreme cases, could result in the cancellation of the project.

The common methods of estimating applied during the design stages can be broadly divided into two major types of technique: *setting (or checking) the budget* or *distributing the budget*. The techniques of estimating to be described in this chapter are given in Table 6.2 with their type or category noted.

A great deal of criticism has been levelled at many of the above techniques because of the static nature in which they express relationships between the variables in a project. Most of them deal with only two variables (quantity and cost) and make little or no attempt to make adjustments beyond the obvious ones of time and function.

Despite these criticisms and drawbacks, most cost planners continue to base their cost predictions on the methods noted above, and for this reason it is important that cost planners are aware of the details of each method, their advantages and disadvantages, and of their suitability through the various stages of a project.

Table 6.2 Categories of Estimating Techniques

Budget Setting Techniques	Budget Distributing Techniques
• Developer's budget	• Approximate quantities
• Functional unit	• Elemental cost planning
• Superficial	
• Functional area	
• Storey enclosure (no longer used)	
• Cube (no longer used)	

Budget Setting Techniques

Developer's budget
The developer's budget or development appraisal approach is usually prepared for guiding the purchase value of a suitable site. This approach is also known as the *residual* analysis and is expressed in the basic equation where the value (or income) generated is compared with the costs of the development:

$$V = L + B + F + P + M$$

where:
V = Value, income or revenue derived from the proposed development;
L = Land costs and associated expenses;
B = Building costs, allowances and fees;
F = Finance costs on land and buildings;
P = Developer's profit (management costs or risk).
M = Marketing costs

This equation determines the *residual* amount, usually land during the early stages of a project. Most commonly, it is used when a decision has to be made on the price to be paid on the land when the project use and broad characteristics can be assessed. The developer's budget can also be used to determine the maximum amount to be spent on the buildings (capital works) if the cost of land is already known. The proposed use and scale of the scheme sets the site value. This appraisal occurs immediately prior to, or during the Briefing Stage.

Functional unit method
This method is also known as the cost per functional unit method. It is one of the simplest and coarsest in application. It consists of selecting a suitable standard functional unit of use for the project, and multiplying the projected number of units by an appropriate cost per functional unit.

The RICS (1969: 10) *Standard Form of Cost Analysis* (see Appendix) as the cost database for much of our cost information has defined functional units as:

> The functional unit shall be expressed as net usable floor area (offices, factories, public houses, etc.) or as a number of units of accommodation (seats in churches, school places, persons per dwelling, etc.).

The AIQS (2000: A8–5) also define functional unit, but do not include net usable floor area (UFA) as part of their definition as follows:

> … the units of performance or occupancy for which a building or section of a building is functionally designed.

Examples of cost per functional unit are:

- Schools, universities, colleges – per student.
- Hotels, hospitals – per bed.
- Offices – per workstation.
- Restaurants, cafeterias – per seat.
- Car parks – per car space.

This method can only be used where the building's use can be analysed consistently in terms of its functional units. Some buildings are difficult, if not impossible, to define a relevant and consistent functional unit. Banks, recreational facilities, renovations/extensions of buildings and factories often defy such functional analysis.

Using this method, the total estimated costs for a new teaching hospital would be:

- Number of functional units: 640 beds
- Cost per bed: £16,000
    Total estimated cost   £10,240,000

Sources of information
Whilst it is probably the best course of action to use your own office or individual project cost information for pricing up any new similar project, in practice you may not have such cost data available. So, we invariably have to rely on the Building Cost Information Service (BCIS) and published sources of building cost information based on the following parameters:

- *Number of functional units*: These are often obtained from the client who may express their initial requirements in such terms. A 1,200 seat cinema, for instance. The design team may also derive this information from early discussions with the client during the briefing stage.

- *Cost per functional unit:* Some professional quantity surveying practices and government offices keep records and analyses of past projects' functional unit costs. The BCIS (1969) Standard Form of Cost Analysis requires this information under the section on *Function and Shape*. See the Appendix for the RICS (1969) *Standard Form of Cost Analysis* for more detailed information.

The most commonly used source for this information is contained in the BCIS and the appropriate sections of the annually or quarterly published building price books such as:

- *BCIS Online* (on demand by subscribers – UK).
- *Spons* (Annual – UK).

- *Laxtons* (Annual – UK).
- *Rawlinsons* (Annual – Australia).

Advice on the selection of the unit method in practice
The unit method can only be used during the very early stages of a
project (Options Appraisal and Strategic Briefing), when the amount of
accommodation/functional units can be determined and where the client
requires a global estimate of costs before any sketches or drawings are
prepared. This is because there is insufficient project detail to apply more
detailed and accurate costing approaches.

Estimates based on this method are very approximate and will vary greatly
depending on the type of construction and standard of finish. These early-
stage estimates are useful for Boards, Trusts, Departments, Councils and
Committees to judge whether their capital expenditure is likely to exceed, be
within, or well below their allocated budgets. The future capital expenditure
programme in an Education or Health Department, for instance, may be
based largely on functional unit methods of estimating.

This form of estimating should not be used beyond the briefing stage of a
project as more accurate methods are available. However, it is commonly
used at later stages in the project, specifically for the building services
elements of the building. For example, the sanitary fixtures, electrical
installation, gas service, water supply and fire protection may be estimated
on a cost per unit until the later stages of the design process. While this
should not occur in modern practice, there are many situations where the
quality of information in the services elements has not been developed at the
same rate as the 'building elements' and has not progressed beyond the level
of counting the number of points or fixtures.

The design information in the critical high cost elements should be advanced
at a rapid rate because of their influence on the total costs. Others may
not progress at the same rate, and consequently, the choice of estimating
technique must reflect the amount and quality of information available. Often,
for simplicity it is assumed that the project information is developed equally
across the elements. However, in practice, for reasons of project characteristics,
complexity and design pressures, information will not be developed equally.
Sophisticated estimating techniques may co-exist with the coarse, rough-and-
ready methods in different elements because the elemental information
varies widely. There is nothing wrong with this, but the cost planner must
not impose an inappropriate technique on an element whose information
database cannot support it. The recommended estimating techniques suitable
for each stage are broad guidelines and should not be considered prescriptive
for that stage, in every element, in every project. Judicious application of
theory as given in these chapters should be tempered with project realities.

A sample of the type of this type of data available on the BCIS website is shown
in Table 6.3 for hotel developments based on March 2006 prices and adjusted to

Table 6.3 Hotel: Functional Unit Prices

Functional Unit Prices							
Functional unit rate for the building excluding external works and contingencies and with preliminaries apportioned by cost. Last updated 1 April 2006. Adjusted to a location index of 96 (Lancashire (Index: 96, Sample: 164)).							
	£/Functional Unit						
Building Function	Mean	Lowest	Lower Quartile	Median	Upper Quartile	Highest	Sample
New Build							
852. Hotels							
m² UFA	1,697	1,294	–	1,603	–	2,288	4
Number of Bedrooms	56,585	17,371	42,084	46,980	68,541	139,543	27

Source: Adapted from BCIS website.

an index of 96 for a Lancashire, UK location. The sample of 47 new build hotels has, however, been taken from the whole cost database in all locations.

Cost per bedroom rates

The functional unit for a hotel type of building is the bedroom and inspection of the cost per bedroom rates on the BCIS database shows the following functional rates based on costs per square metre of UFA (usable floor area) and costs per bedroom. Since the sample of four projects on the costs per square metre UFA is low, it is not recommended to use this as a sound basis for costing the new project. The sample of 27 for costs per bedroom is a much broader base for cost analysis and these costs per bedroom of £56,585 (Mean), £46,980 (Median) with a range from £17,371 to £139,543.

The location and tender price index (TPI) dates shown in Table 6.3 have been automatically adjusted by the *BCIS Online* system and these are shown to make the user aware of the basis that the cost information is presented. So, in our case we can see that the location is based on a national location index of 96 for Lancashire and an index automatic adjustment to the 1 April 2006. See Chapter 5 for a discussion of the types and use of the BCIS indices.

Question 6.1
Functional Unit Budgets

Calculate the costs of the following buildings based using appropriate rates taken from the BCIS Table 6.3 as the source of cost data for preparing the estimate.

(a) Hotel
Suburban single or double storey – limited accommodation with toilet facilities, large bar areas, partly air-conditioned.

The mean of £57,000 has been chosen as the basis for pricing in this instance. The BCIS is one of the best public sources of functional unit information. Many private sources exist in private practice and government.

Cost per bedroom unit: £57,000 per unit (Manchester) taken from data given in the BCIS, March 2006.

Number of units: 95 bedrooms

(b) Secondary school
Using the functional unit data from the BCIS shown in Table 6.4 prepare a budget estimate for the secondary school (high school) located in Southampton.

We are going to assume that the rates indicated in the BCIS for Southampton shown in Table 6.4 would be satisfactory in this situation. Thus, a rate of £7,430 per pupil is used. This is rounded up from £7,423 shown as the mean cost per pupil in the Table.

Number of functional units (students): 450 pupils

The total cost or budget for the buildings only to be calculated.

Cost per pupil rates

An extract from the BCIS functional unit price section for schools is shown in Table 6.4. As noted in Table 6.3 the rates have been adjusted for the Southampton location (Hampshire), which has a location index of 107 and time indexed to a date of 1 April 2006.

Superficial method
This method is also known as the floor areas method. It is similar to the unit method and uses a single rate technique (cost per square metre) of estimating to calculate the cost of a building. The superficial method is one of the most popular methods of estimating and is commonly used during the early stages of a project. By necessity, the method can only be used when the floor area of the building can be assessed or measured reasonably accurately from drawings; even if only in an outline form.

It is a simple technique where the GFA of the building in square metres is multiplied by a suitable rate per square metre. The GFA and the net floor area of a building are defined by the BCIS (1969: 9). See the Appendix for the RICS (1969) *Standard Form of Cost Analysis* for more detailed information.

Table 6.4 Schools: Functional Unit Prices

Functional Unit Prices							
Functional unit rate for the building excluding external works and contingencies and with preliminaries apportioned by cost. Last updated 1 April 2006. Adjusted to a location index of 107 (Hampshire (Index: 107, Sample: 289)).							
Building Function	£/Functional Unit						Sample
	Mean	Lowest	Lower Quartile	Median	Upper Quartile	Highest	
New Build							
712. Primary Schools							
m² UFA	1,707	809	1,243	1,649	1,984	5,894	122
Number of places	5,735	1,649	3,853	5,340	6,970	23,290	293
713. Secondary (High) Schools							
m² UFA	1,363	765	1,168	1,325	1,564	2,140	31
Number of places	7,423	703	3,697	6,353	9,117	29,028	54
713.1 Secondary Schools: Specialised Teaching Blocks							
m² UFA	1,563	808	1,128	1,381	1,822	3,555	45
Number of places	7,065	2,485	3,958	6,063	8,424	16,926	26
713.8 Secondary Schools: Mixed Facilities							
m² UFA	1,810	1,467	–	1,839	–	2,095	4
Number of places	7,171	2,382	–	6,430	–	12,702	3

Source: Adapted from the BCIS website.

Example
Using the superficial method calculate the costs for an office building project with the following data:

- Area of site   600 m²

- Plot ratio   2.5:1

- Total gross area of building permitted
  on this site = 2.5 × 600 m²   = 1,500 m²

- Cost per square metre   £1,200

  Total estimated cost   £1,800,000

Sources of information

- Total area of building
  As noted in the example above the GFA may be derived from planning scheme data. The assumption being made is that the client and the design team will produce a project with a floor area to the limit of what is permitted by the planning authority. Similarly, where the planning scheme defines the number of storeys or the maximum floor area (as many do) this provides essential data for the superficial method of estimating and it requires no drawings and can be carried out during the early stages of Establishing the Need and Options Appraisal.

  When drawings or sketches have been prepared (Outline Proposals or Detailed Design) the area is obtained from those drawings.

  Sometimes, at the Briefing stage, a client may define his/her requirements in terms of total floor area, or more commonly in terms of Net Floor Area (see definitions in BCIS, 1969: 9–10). See the Appendix for the RICS (1969) *Standard Form of Cost Analysis* for more detailed information. The terms, net usable floor area, net lettable area (NLA) or net rentable area (NRA) may be used by property analysts. In residential buildings the term, Net Habitable Floor Area is often used.

  For Australian definitions of other commonly used terms see AIQS (2000) and Property Council of Australia (1997) *Method of Measurement*.

  In the Strategic Briefing stage where information is sparse, the superficial method may be used, but ensure that due allowance is made for Circulation space, Ancillary space and Internal Divisions (see BCIS, 1969: 9 for definitions). These areas are often frequently referred to generically as *circulation* spaces and may represent between 10% and 30% of the total floor area (GFA).

Using the BCIS as our source again in Table 6.5 we can identify the costs per square metre of the hotels and secondary schools used earlier in our questions. Again, the location and TPI dates have been automatically adjusted by the *BCIS Online* system. So, in our case we can see that the location is based on a national location index of 96 for Lancashire and an index automatic adjustment to the 1 April 2006.

The questions have now been revised to include the GFA of each scheme.

Based on a sample of 47 new build hotels in all UK locations a mean cost per square metre of £1,119 has been calculated and a median cost of £1,092/m^2. The range of costs is wide from a low £652 to £2,120/m^2; probably representing the wide range of quality from budget hotels to five-star rated.

Table 6.5 Hotels: £/m² Study

Cost Per Square Metre Study							
Rate per m² gross internal floor area for the building excluding external works and contingencies and with preliminaries apportioned by cost. Last updated 1 April 2006. Adjusted to a location index of 96 (Lancashire (Index: 96, Sample: 164)).							
	£/m² Gross Internal Floor Area						
Building Function	Mean	Lowest	Lower Quartile	Median	Upper Quartile	Highest	Sample
New Build 852. Hotels	1,119	652	906	1,092	1,282	2,120	47

Source: Adapted from the BCIS Website.

Question 6.2
Superficial Budgets

Calculate the costs of the following buildings based using appropriate rates per square metre taken from the BCIS tables shown as the source of cost data for preparing the estimate.

(a) Hotel
Suburban single or double storey – limited accommodation with toilet facilities, large bar areas, partly air-conditioned.

The mean of £1,120/m² has been chosen as the basis for pricing in this instance.

GFA: 5,200 m²

(b) Secondary school
Using the functional unit data from the BCIS shown in Table 6.6 prepare a budget estimate for the secondary school (high school) located in Southampton.

This information may be available in one of the published sources noted above. However, for our purposes we are going to assume that the rates indicated in the BCIS for Southampton shown below would be satisfactory in this situation. Thus, a mean rate of £1,060/m² is used and taken from Table 6.6, Secondary Schools, mean value = £1,059, say round up to £1,060.

Estimated GFA: 3,600 m²

The total cost or budget for the buildings only to be calculated.

Table 6.6 Schools in Hampshire: £/m² Study

Cost Per Square Metre Study							
Rate per m² gross internal floor area for the building excluding external works and contingencies and with preliminaries apportioned by cost. Last updated 1 April 2006. Adjusted to a location index of 107 (Hampshire (Index: 107, Sample: 289)).							
Building Function		£/m² Gross Internal Floor Area					
	Mean	Lowest	Lower Quartile	Median	Upper Quartile	Highest	Sample
New Build							
710. Schools							
Generally	1,179	464	937	1,133	1,369	3,346	1,552
Public	1,176	464	937	1,133	1,366	3,346	1,391
Private	1,226	472	964	1,178	1,426	2,629	151
713. Secondary (high) Schools							
£/m²	1,059	541	840	1,004	1,226	3,346	240
713.1 Secondary Schools: Specialised Teaching Blocks							
£/m²	1,235	510	1,007	1,179	1,428	2,323	202

Source: Adapted from the BCIS Website.

Question 6.3
New Offices in Liverpool

A client has defined her needs for a new city location in Liverpool for a 10-storey office building in terms of 1,260 m² of NRA. The project will have air-conditioning, lifts, high-quality finish and no car parking. Prepare an estimate of the cost of the building, exclusive of external works, car parking and other allowances.

The present date cost per square metre of GFA for similar buildings has been assessed at £1,200. Calculate the present date cost of this project. In similar situations you may source relevant cost information published in the BCIS or published sources.

Further consideration on the application of these budget setting techniques

The area used in the superficial method is the *gross floor area* (GFA), which is defined by the BCIS (1969: 9) as:

> ... the total of all enclosed spaces fulfilling the functional requirements of the building measured to the internal structural face of the enclosing walls.

Measurement of the GFA (RICS, 1969: 9) should *include*:

- All area occupied by partitions, columns, chimney breasts, internal structural or party walls, stairwells, lift wells and the like.
- Lift, plant and tank rooms and the like above the main roof slab.
- Sloping surfaces such as staircases, galleries, tiered terraces and the like should be measured flat on plan.

Measurement of the GFA (RICS, 1969: 9) should *exclude* the following unenclosed spaces, which should be shown separately:

- Any spaces fulfilling the functional requirements of the building, which are not enclosed spaces (e.g. open ground floors, open covered ways and the like).
- Private balconies and private verandahs.

Contrast this UK approach with the Australian approach where the GFA is 'the sum of the *fully enclosed covered area (FECA)* and *unenclosed covered area (UCA)*' (AIQS, 2000: A8-1 to A8-10). These two areas are important to cost planning practice in Australia as they reflect design practice (and climate imperatives) where UCAs can be a significant proportion of the total area of many projects. These areas are used throughout the design stages, being clearly documented on the standard forms (AIQS, 2000) and defined as follows:

Fully enclosed covered area (FECA)

The sum of all such areas at all building floor levels, including basements (except unexcavated portions), floored roofing spaces and attics, garages, penthouses, enclosed porches and attached enclosed covered ways alongside buildings, equipment rooms, lift shafts, vertical ducts, staircases and any other fully enclosed spaces and usable areas of the building, computed by measuring from the normal *inside* face of the exterior walls but ignoring any projections such as plinths, columns, piers and the like which project from the normal inside face of exterior walls. It shall not include open courts, light wells, connecting or isolated covered ways and net open areas of open portions of rooms, lobbies, halls, interstitial spaces and the like which extend through the storey being computed.

Unenclosed covered area (UCA)

The sum of all such areas at all building floor levels, including roofed balconies, open verandahs, porches and porticos, attached open covered ways alongside buildings, undercrofts and usable space under buildings, unenclosed access galleries (including ground floor) and any other trafficable covered areas of the building which are not totally enclosed by full height walls, computed by measuring the area between the enclosing walls or balustrade (i.e. from the inside face of the UCA excluding the wall or balustrade thickness). When the covering element (i.e. roof or upper floor) is supported by columns, is cantilevered or is suspended, or any combination of these, the measurements shall be taken to the edge of the paving or to the edge of the cover, whichever is the lesser. UCA shall not include eaves overhangs, sun shading, awnings and the like where these do not relate to clearly defined trafficable covered areas, nor shall it include connecting or isolated covered ways (element XB).

The rate or cost per square metre used in the estimate is intended as a rate for the cost of a standard or typical building within that function/quality range. The cost of external works, alterations, abnormal items such as piling or unusual foundations, locality and contingency allowances are normally allowed for separately as lump sums or percentages of the Building Cost and are added to the superficial area calculation.

Example
An office block is proposed in Guildford and the developer's budget indicates office space with a NRA of 12,000 m^2 is required. What would be the GFA requirements when an efficiency of 75% is expected?

$$\text{GFA} - 12,000\,\text{m}^2/0.75 = 16,000\,\text{m}^2$$

So, the building costs for the office block are based on a superficial area of 16,000 m^2 GFA and not a NRA of 12,000 m^2.

Sources of information for superficial rates:

- Internal office records
  See the Appendix for the RICS (1969) *Standard Form of Cost Analysis* for more detailed information.
  As noted above all costs are expressed in cost per square metre (GFA).

- The BCIS
  A typical analysis of rates per square metre for offices from the BCIS is included in Table 6.7 to indicate the type and range of information available for cost planning purposes. The location and TPI dates have been automatically adjusted by the *BCIS Online* system. The location is based on a UK mean location and a time index automatic adjustment to the 4th quarter 2005.

- Published cost guides as noted earlier:

Spons:	Part IV approximate estimating: building prices per square metre.
Laxtons:	Square metre prices.

Table 6.7 Offices: Costs Per Square Metre Study

Cost Per Square Metre Study							
Rate per m² gross internal floor area for the building excluding external works and contingencies and with preliminaries apportioned by cost. Last updated 18 March 2006. At 4Q2005 prices (based on a TPI of 223) and UK mean location.							
Building Function	£/m² Gross Internal Floor Area						Sample
	Mean	Lowest	Lower Quartile	Median	Upper Quartile	Highest	
New Build 320. Offices							
Generally	1,207	353	899	1,122	1,401	3,504	795
Steel Framed	1,179	353	914	1,124	1,339	3504	334
Concrete Framed	1,374	637	1,034	1,319	1,627	3,135	267
Brick Construction	1,017	518	790	926	1,180	2,604	168
Timber Framed	922	671	772	905	1,074	1,170	13
Air-Conditioned	1,374	518	1,054	1,287	1,593	3,504	281
Not Air-Conditioned	1,038	521	806	983	1,198	2,400	269
1–2 Storey	1,033	353	800	970	1,207	2,604	346
3–5 Storey	1,237	521	969	1,168	1,401	3,504	307
6+ Storey	1,657	660	1,308	1,584	1,998	3,135	122

Source: Adapted from the BCIS Website.

Rawlinsons:	Estimating – building costs per square metre.
Cordells:	Building cost indicators.
Building Economist:	Building costs per square metre.

These publications provide sophisticated and detailed costs for various types and different quality buildings. In selecting a rate try to ensure that it is based on a similar quality building, same function, of comparable quality and storey heights and spans. Where a building consists of parts, which are of different types, functions and quality, then each part should be measured separately and different, but appropriate rates applied to each part. Interpolation of rates is commonly practised where a single rate cannot be used on the proposed project.

Question 6.4
Offices in Slough

Prepare an estimate for multi-storey (4-storey) offices in the city of Slough including air-conditioning.

GFA $= 1,780\,m^2$

Allow for external works and surface car parking – estimate provided £252,250.

Rate per square metre – to be obtained from the BCIS as noted above or any one of the other published cost guides listed above.

Advice on the selection of the superficial method in practice

As with the unit method this technique can only be used during the early stages of a project. That is, during the early stages of Options Appraisal and Strategic Briefing. The project must have progressed beyond expressing the requirements in terms of units and a reasonably accurate assessment of the floor area of the proposed project must be able to be measured (possibly from preliminary sketch drawings), or alternatively, be appraised from the client's statement of requirements, the planning scheme or by being given explicitly in square metres. This method, as with all these early-stage techniques, must not be used when better quality and more accurate information becomes available as more reliable and accurate techniques can be used as will be explained later.

Advantages

- The unit of measurement (GFA) is generally meaningful to all the parties – especially the client and the architect.

- Most published cost data is expressed in costs per square metre.

- Estimates are prepared quickly and are simple, but beware of the method's inherent simplicity. The accuracy is conditional upon choosing a rate from a project with similar characteristics.

Disadvantages

- Difficult to make adjustments to the estimate for differences in plan shape, storey height, site conditions, construction techniques, quality and quantity. All of these can have major cost implications as noted in design economics in Chapter 3.

- On a practical note, always check whether the estimate calculation is for building(s) only. Most of the published building price books indicate whether they include amounts for major service installations such as lifts, air-conditioning and fire protection. Separate and additional allowances often have to be made for items considered to be above the average or stated standard and for work beyond the external face of the building (or within 3.0 m of it). Such items excluded from the cost per square metre GFA may include external landscaping, external services, special earthworks and foundations.

Functional area method
Earlier in this chapter, the unit (or functional unit) method of estimating was described and is based on the broadest or coarsest level of costing based on the cost per functional unit or functional place. Little published work is available from the UK, but in Australia the functional space was developed in the late 1970s for schools projects. This approach had the support of the Commonwealth (central) government (1978) in a major cost study of schools in Australia. The study pioneered the use of functional area/cost analysis.

The basis for costing is not the GFA as in the superficial method, but a refinement of this overall area into the component functional spaces. So, once the type and area of space could be calculated this method can be used. The method provides a bridge between the coarseness of the Strategic Briefing stage and the development of the brief into the Outline Proposals stage. The Functional Area method of estimating combines or integrates the functional unit method with superficial method of estimating. It refines both methods to provide a technique based on the costing of functional areas with appropriate rates per square metre, or costs per functional unit.

Functional area is defined by the AIQS (2000: 5–7) *Australian Cost Management Manual* as:

> Any group of accommodation, which has a common work function within a particular type of building. It includes all circulation necessary within that area.

The objective '… of functional analysis is to derive functional areas and costs for use in budgeting' (AIQS, 2000: 5–4). It is useful to quote direct from the AIQS (2000: 5–4 and 5–5) to gain a better understanding of the background to the technique.

> Functional analysis is essentially a budgeting tool. By clearly defining the various space requirements of a building and (its) costs … provides greater flexibility and accuracy to cost budget procedures, and identifies those functional areas of different quality and thereby cost. The resultant system provides a basis for rapid estimates over a wide range of capital works requirements through the application of functional costs to the Client Authority's accommodation requirements …

There are two ways in which functional analysis may be used in budgeting. The first is to identify the functional areas from the brief and to price these using rates developed from previous functional cost analyses. As architectural briefs normally set out the basic accommodation areas required for the various activities within buildings, it is necessary to determine and apply an accurate percentage allowance (called *Grossing Factor*) in order to calculate the necessary allowances for circulation, travel and engineering areas which, when added to the basic accommodation area requirements, give the GFA

The second method requires the prior rationalisation of reasonable space requirements and costs for each functional area and the establishment of functional unit rates where appropriate, and involves the application of these standards to the Client Authority's requirements. This method, although requiring extensive analysis and rationalisation of historical data to ensure reliability of information, is by far the more effective means of setting budgets.

In AIQS (2000), a list of functional areas (Appendix A, Part 7) is presented for the following building types:

- Primary schools

- High schools/secondary colleges

- TAFE (technical and further education) colleges

- General hospitals.

This document attempts to develop uniformity of identification and naming of functional areas, so that users, designers and cost planners can communicate their requirements and solutions in a common language.

The method of costing relies on a significant investment of time and other resources into analysing a sound database of typical projects that can provide the foundation for future cost planning. For instance, if functional costs need to be developed for high schools, then a representative sample of past projects must be selected to form the basis of the analysis. It is suggested that at least five projects as a minimum are analysed. Consideration and selection of a suitable sample of projects must be a joint and co-operative exercise between the client body, users and design team members who have had some ongoing experience with the particular project type.

An excellent example of this process in action is contained in Commonwealth of Australia (1978) *Comparative Capital Costs of Government and Non-Government Schools in Australia*. It is a model for similar studies and provided a wealth of comparative functional area cost information for the commonwealth and state education departments to plan and cost their future capital works programmes in education.

In terms of creating the cost database for this technique the first stage is to prepare several functional areas cost analyses on typical projects. An elemental cost analysis (based on the actual tender) of the project must be prepared, and the elemental costs are then subdivided between the various functional spaces as defined in the AIQS (2000: Section 5.2). This process is called functional area/cost analysis.

Essentially, at the end of this demanding research and analysis activity, the cost planner will have access to a listing of functional spaces costed at appropriate rates per square metre. These rates can then be used to estimate the cost of planned projects linked to the Strategic Briefing and Outline Proposals stage when functional spaces are being defined. A sample of this type of data is given in Commonwealth of Australia (1978: 45, 46, 51, 52, 65–67, 73 and 74).

Question 6.5
Functional Area costing

The functional areas and present date costs of a typical primary school are given below.

Functional Spaces	Cost £/m^2	Area of New School GFA
1 Library Resource Centre		
Main Reading Area	376	55
General Workroom/Store	464	45
Circulation	nil	5
2 Learning Areas		
Classrooms	363	386
Practical Activities	473	465
Withdrawal Room	445	74
Circulation	nil	19
3 Administration/Staff		
Principal's Office	369	18
Assistant Principal Office	360	10
Reception/Waiting Area	326	12
Store Rooms	420	13
Staff Lounge	353	36
Staff Toilets	736	12
Student Rest Room/Toilet	650	10
Cleaner's Store	757	4
Circulation	nil	8

4 Pupil Amenities		
Pupil Toilets	665	52
Circulation	nil	3
5 Travel/Engineering	106	410

Using the functional area costs given above prepare an estimate of the present date cost of the new primary school.

Notes: (a) Assume the rates per square metre are based on the projected tender date and no further cost adjustments have to be made for time and other factors.

(b) Allowances for contingencies, price and design risk, external works and any other abnormals are *not* included in this estimate.

Advice on the selection of the functional area method in practice

- The functional cost method of estimating provides timely and fitting cost advice to the design team at an early stage of the design process that closely matches the development of information in the design brief. Decisions on the type and extent of functional space in the preliminary project design are reflected in the more accurate costs that this technique brings to these activities.

  The technique provides the opportunity for the cost planner to become more intimately involved in design decision-making during the early crucial stages when the shape, extent and form of the project are still fluid. The technique thereby promotes greater interaction and integration with the client and between design team members.

  The use of this technique in cost budgeting is comprehensively described in the AIQS (2000), *Australian Cost Management Manual*.

- The technique is time and data-intensive in its development stages. It requires a strong commitment on the part of the client body, the design team and the cost planners to collect, analyse and present data in a suitable form for future projects. Such a commitment can only be warranted where the client body is involved in a continuous and consistent programme of building projects. This explains the interest in this technique by bodies such as education departments and health departments, who have large annual capital works programmes containing projects of a similar nature. Some cost planners have also applied this technique to shopping and large retail centres, where the developer(s) have a consistent design approach.

Unless the considerable investment of time in functional area cost analysis can show benefits in costing several projects, then this technique is not recommended for the *one-off* or individual building project. This technique could have more universal application as experience in Australia has shown it to be a reliable and useful tool for use by the cost planner at the briefing stage where functional spaces are being decided.

Storey enclosure method

This method, developed in the early 1950s, has had very little application in practice. It is noted here for reference purposes rather than for practical application.

The storey enclosure method is a single price rate method in a similar form to the unit and superficial method. However, its distinguishing feature is that it was one of the first techniques that attempted to incorporate the cost effect of design features such as the height and shape of the project. Unfortunately, the technique had little opportunity to gain wide acceptance because it was superseded by elemental estimating (see later). No cost data was collected to any significant degree. Therefore, the lifeblood for any technique was missing.

Rules of measurement

Areas of the various floors, roof and external walls are calculated, and each is weighted by a different factor to provide a total number of *storey enclosure units*. To arrive at the total estimate figure, the calculated storey enclosure units are multiplied by a storey enclosure rate obtained from previously analysed projects.

Why did the storey enclosure technique fall into disuse?

- The technique involves far more calculation than either the superficial or cube method.

- No rates for this method are published. Therefore, the cost planner would need to spend considerable time in calculating suitable rates from past projects.

- The technique does not relate client requirements to the actual design. Nor does it assist in guiding cost planning decisions. It was superseded by the elemental method, which was developed at around the same time in the 1950s. See description of elemental method later in this chapter.

- As with most single price rate estimating techniques it is difficult to assess the overall cost effect to changes in specification.

For an interesting comparison of the cube, square metre and storey enclosure methods applied to a single project, see Ferry et al. (1999: 115, 116).

Cube method (cost per cubic metre)
Again, this is a single rate method of estimating, similar to the superficial method, but based on the cubic content of a building rather than the floor area. It is sometimes referred to as the volume method and is not as commonly used as the superficial method, which has superseded it in the UK and Australia. However, whilst many people believe that the technique is used by the USA, Mann (1992: 16) makes the following comment:

> This method, using cubic metres, is more common in the United Kingdom and Europe than in the United States.

Since all parties seem to agree they do not use the method we believe it is probably safe to say the cubic method is dead! In Australia and the UK it is certainly obsolete.

For interest's sake or just in case you wish to test the technique, the rules of estimating are simple. The number of cubic metres, or cubic content of a building is multiplied by an appropriate rate to give the estimated cost. If used, this method would only be used during the early stages of a project; soon after outline proposals or drawings/sketches have been prepared.

Example
An office block with a cubic content of 16,000 m^3 has been estimated to cost £325.00/m^3.

- Cubic content   16,000 m^3
- Rate per cubic metre   £325.00/m^3

  Total estimated cost   £5,200,000

Rules of measurement
The cubic content is normally calculated from measuring the length and breadth of a building between the normal *outside* face of the external or enclosing walls. The height of the building is measured from the top of the foundations to half-way up the slope of a pitched roof; or in the case of a flat roof, to 0.61 m above the level of the roof. The volumes of any projections (porches, dormers, chimney stacks, ducts, etc.) are added to the volume previously calculated.

As in the superficial methods the costs of external works, abnormal items such as special foundations, locality and contingency allowances must be added separately to the calculated cubic estimate.

Sources of information
Except for a few offices, which still calculate their own cubic rates, there are no published sources of information for this cost data. For this reason this method is generally never used.

The only areas where this method is used today, is where the cubic content has a bearing on the design of a particular element such as heating/cooling and ventilation requirements, and sometimes in the calculation of fire insurance premiums. It is noteworthy that the BCIS (1969) Standard Form of Cost Analysis retains the cubic metre of treated space as an element unit quantity for two elements (5F Space Heating and Air-conditioning and 5G Ventilation).

The rules of measurement for internal cubic quantities in these two elements is given in BCIS (1969: 10) where the cubic content:

> ... shall be measured as the net floor area of that part treated, multiplied by the height from floor finish to the underside of the ceiling finish (abbreviated to Tm^3).

In contrast, the AIQS (2000) requires that heating, ventilation, cooling and air-conditioning areas are measured in square metres as the *treated area* for each engineering service.

Advice on the selection of the cubic method in practice
As with the previous methods, functional unit and superficial, the cubic metre method can only used when the cubic content of the building can be reasonably calculated and this may be after the Strategic Briefing stage, when the building volume information available is limited. However, its application is limited due to lack of cost data, as noted earlier.

Advantages

- Simple, quick and easy to use.

- Satisfactory on works of a recurring nature.

- Suitable where cubic content of a space or a building directly influences the cost of an element (see the two elements noted above).

Disadvantages

- Clients, architects and most design team members tend to relate to floor area rather than volume.

- Adjustments for plan shape, quantity or number of floors and quality are not easily made.

- No published sources of cost data.

- The cubic rates are sensitive because such large volumes are involved. It may be necessary to express the rates to three or four places of decimals.

Reflection
It is essential that we place our techniques in the pre-design and design periods in context. The practice of cost planning takes place at our six plus one RIBA (2000) Plan of Work design stages. These stages take us from the inception of a project where we establish the need for the project to the production information and tender stage. Before we embark on a detailed description of cost planning practice, it is well to remind ourselves of the stages involved in the whole pre-construction process. These stages are shown in Figure 6.3, together with the various budget setting (described in this chapter) and budget distributing techniques to be described in later chapters.

Some of the early-stage techniques that lie outside the scope of our cost planning focus are noted in Figure 6.3. The importation of existing techniques

Pre-design			Design			
Inception or Feasibility			Pre-construction Period			
Pre-stage A	Work Stage A	Work Stage B	Work Stage C	Work Stage D	Work Stage E	Work Stage F
	1	2	3	4	5	6
Establishing the Need	Options Appraisal	Strategic Briefing	Outline Proposals	Detailed Proposals	Final Proposals	Production Information
EARLY STAGE TECHNIQUES (Described later in Chapter) • Strategic Value Management • Strategic Needs Analysis • Quality Function Deployment • Situation Structuring		Cost Limit	Cost Targets		Cost Checking	
	• *Expert Choice*					
	BUDGET SETTING TECHNIQUES		**BUDGET DISTRIBUTING TECHNIQUES: BUDGET DISTRIBUTION, CHECKING AND RECONCILIATION**			
	• Developer's Budget					
	• Total cost					
	• Functional Unit (student, bed, seat, room)					
		• Superficial (gross floor area)				
		• Functional Area/Cost				
			• Elemental (group)			
				• Elemental (individual and sub-elemental)		
			• Approximate Quantities (to suit level of detail available)			

Figure 6.3 Cost Planning Techniques and the Outline RIBA Plan of Work

such as Strategic Needs Analysis (Smith, 2005), *Expert Choice* or the analytical hierarchy process (Saaty, 1990a, 1990b; Yang and Lee, 1997), Quality Function Deployment (QFD) (Akao, 1990 and Kamara et al., 1999), Analytical Design Planning Technique (*ADePT*) (Austin et al., 2000), *Situation Structuring* (Dickey, 1995; Wyatt, 1999) value management and strategic value management (Kelly and Male, 1993 and Thiry, 1997) is likely to expand in the future and the reader is advised to refer to the noted texts to improve their knowledge in this growing area.

The techniques applied to the early stages of the project life reviewed in this chapter and the use of the traditional cost planning techniques described in later chapters are summarised in Figure 6.3. This Figure suggests the use and timing of the various budget setting approaches in the pre-design period and indicates when the budget distributing cost planning techniques are used in the remaining design or pre-construction stages.

Many of the early-stage techniques are simple to use, especially the single rate methods described. The techniques become more complex as the project information base expands. The essential skill required in producing an accurate estimate is not in carrying out the necessary calculations, which are very basic. It lies in applying the mixture of skill, experience and judgement to selecting an appropriate rate for the assessed, assumed or perceived quality of the proposed project. In addition, the inclusion of the various additional allowances to be appended to the net building costs requires a broad understanding of the total project requirements. In many cases these additions can represent a range from 20% to 50% of the total building cost.

It is essential for the cost planner to be mindful of the characteristics of the various techniques; be familiar with the stage best suited for the application of the technique; and to appreciate the advantages and drawbacks to their use. A model hierarchy of use for these various techniques through the different stages may be useful to provide a guide to the cost planner. However, the cost planner must be conscious of the range of techniques described here and recognise that techniques can be used in tandem at the same stage for different parts of the project. The determining factor in their use is not the theoretical or recommended stage for their application, but the type and quality of information suitable for use in the particular technique.

It is also necessary to emphasise that the preparation of an estimate using any of these techniques is unlikely to be a one-off procedure. A number of estimates may have to be prepared and presented to the design team and the client. A review of the estimate(s), based on feedback from these meetings when quality, quantity and other factors must be discussed, can be incorporated into the revised scheme and into the estimates.

Question 6.6
Cost Planning Techniques

A number of methods of preparing estimates for cost plans have been described to this point in this chapter.

Summarise the main features of these methods. Taking account of the information available, indicate the stage in the RIBA design period to which they are most suited.

COST PLANNING IN THE STRATEGIC BRIEFING STAGE A

Design teams are usually required to produce a first estimate at the Brief stage, that is, long before any drawings have been prepared or any specification has been decided. Clients expect to be able to make the decision to proceed or abandon the project at the end of the Strategic Briefing stage. So they need to have reasonably accurate estimates from the design team if they are to avoid the risk of serious or even catastrophic difficulties later on.

Information Usually Available at this Stage

The initial strategic brief (or statement of need) is usually the only information available at this stage. It provides short statements on the major factors of type and area of space likely to be needed; an indication of quality in general terms and any site information.

As noted earlier (RIBA, 2000) strategic briefs usually only contain concise descriptions of:

- Needs defined both emotional and physical.
- Concept description.
- Objectives of the project.
- *Use* to which the building is to be put (e.g. hotel, school, offices) and major functional relationships.
- *Floor area* required (to which additions may have to be made to allow for circulation areas, plant rooms, etc.).
- An indication of the *quality* required (e.g. prestige v. average v. low).
- Planning, environmental criteria.
- Initial cost/time criteria (if known).
- Basic details of the site.

To illustrate the limited nature of the information that has a bearing on the project costs in an initial Strategic Brief a typical example is given in Table 6.8 for a new office building

Table 6.8 Example of Initial Brief for an Office Building

Typical Information	
	New Offices
• **Space**	A NLA of 1,000 m² and parking for 20 cars.
• **Use**	Premises will be offered for leasing as normal office accommodation.
	Kitchen and dining accommodation is required for a total of 70 people.
• **Quality**	Average commercial quality.
• **Location**	Southampton, suburban site (not city centre).
• **Site**	The site is cleared of houses previously occupying the site.
	A site plan showing details of the site boundary, dimensions, adjoining properties and streets may be available. A copy of the title may also be obtained to obtain this useful information.

Interrogation of the BCIS

The cost planner has to gain relevant cost information that will provide the necessary data to begin the process of familiarisation with this type of project. From the initial information provided in the outline brief we can commence the process of gaining information from the BCIS that will assist in this process.

The key data or keywords from the outline brief can be summarised as follows:

- A primary function of new offices (new build).
- Lettable area of 1,000 m², with a need to adjust to derive the GFA.
- An indicative number of 70 persons occupying the offices.
- New build for average commercial leasing rather than owner occupation.
- Location on the south coast, UK, on a suburban site.
- Site is already cleared, but may be a need to make provision for possible additional cost of foundations or groundworks.

As noted in the previous chapter the BCIS provides us with the capability of gaining cost data at various levels of detail to match the project information available. In our case we shall begin at the broadest level of analysis, that is, the functional unit prices and the cost per square metre (GFA).

Recognising the primary function as offices and new build we can gain this cost data in the form as shown in Table 6.9.

Table 6.9 Functional Unit Prices – New Build: Offices

Functional Unit Prices							
Functional unit rate for the building excluding external works and contingencies and with preliminaries apportioned by cost. Last updated 18 March 2006. At 4Q2005 prices (based on a TPI of 223) and UK mean location.							
	£/Functional Unit						
Building Function	Mean	Lowest	Lower Quartile	Median	Upper Quartile	Highest	Sample
New Build							
320. Offices							
m² UFA	1,480	616	1,132	1,376	1,683	3,687	166
m² Net lettable floor area	1,860	678	1,300	1,781	2,135	3,665	22
Number of persons	24,398	13,004	15,309	18,562	25,148	64,313	8

Source: Adapted from the BCIS Website.

Inspection of this cost data indexed to the 4th quarter 2005 in a UK mean location (not a specific region) starts to build up the *picture* we need to be able to advise the design team of a suitable cost limit or budget for the proposed scheme. When we review the costs here, there is a wide range from lowest to highest costs. For instance, in cost per square metre of NLA based on a sample of 22 projects, the costs range from £1,860 to nearly double at £3,665. A similar trend is shown for costs per person, but this sample is only eight projects and the functional unit for offices is not as consistent as it is for other types of function such as schools and hospitals.

Next, we shall inspect the costs per square metre GFA, a more consistent form of analysis. The costs for offices and new build are shown in Table 6.10, which is Table 6.6 repeated for ease of use.

The costs are again based on the 4th quarter 2005 and a UK mean location and provide cost data in a similar format to that in Table 6.9 earlier, except that we now have more detailed cost information on the type of structures of these new build offices, air-conditioned and non-air-conditioned offices and how the number of storeys affects the overall cost per square metre GFA. For

Table 6.10 Costs Per Square Metre: New Build Offices

Cost Per Square Metre Study							
Rate per m² gross internal floor area for the building excluding external works and contingencies and with preliminaries apportioned by cost. Last updated 18 March 2006. At 4Q2005 prices (based on a TPI of 223) and UK mean location.							
	£/m² Gross Internal Floor Area						
Building Function	Mean	Lowest	Lower Quartile	Median	Upper Quartile	Highest	Sample
New Build 320. Offices							
Generally	1,207	353	899	1,122	1,401	3,504	795
Steel Framed	1,179	353	914	1,124	1,339	3,504	334
Concrete Framed	1,374	637	1,034	1,319	1,627	3,135	267
Brick Construction	1,017	518	790	926	1,180	2,604	168
Timber Framed	922	671	772	905	1,074	1,170	13
Air-Conditioned	1,374	518	1,054	1,287	1,593	3,504	281
Not Air-Conditioned	1,038	521	806	983	1,198	2,400	269
1–2 Storey	1,033	353	800	970	1,207	2,604	346
3–5 Storey	1,237	521	969	1,168	1,401	3,504	307
6+ Storey	1,657	660	1,308	1,584	1,998	3,135	122

instance, in reviewing this cost data we can expect that the mean cost per square metre of new build offices is £1,207. This cost can fluctuate between £353 at its lowest and £3,504 at its highest in this large sample of 795 offices. Brick construction also appears to be cheaper than either steel or concrete framed construction. Timber framed is cheaper still, but is a less popular form of construction with only 13 projects sampled. Whether the offices are air-conditioned or not also appear to be an important design factor because the costs per square metre of air-conditioned offices vary from 30% to 45% higher in costs when we inspect these analyses of over 500 projects (281 + 269 = 550). A similar pattern emerges when we review the variation in costs due to the number of storeys. Low-rise offices of 1 or 2 storeys with a mean costs per square metre of £1,033 are nearly 40% lower in cost than the high rise of 6 storeys and above offices with a mean of £1,657.

With this data interrogation of the BCIS has given us some important pointers to the development of this office project. Whilst we can globally estimate the cost of these offices relatively easily the design team and the client should provide early decisions on three significant factors that influence the costs per square metre of our project and highlighted by Table 6.10. These factors are:

- The efficiency of the building in terms of NLA to GFA (percentage of circulation space).

- Number of storeys.

- Is the building air-conditioned or not?

Initial Costing

Using the data we have from the BCIS and from the outline brief, we can begin to identify a budget or range of costs for our proposed new build offices. We shall use the mean costs as presented in Tables 6.9 and 6.10. Table 6.11 summarises the results.

Table 6.11 Budget Calculation

Functional Unit	Quantity × Rate	Total Cost	Index
• NLA	1,000 m² @ £1,860	£1,860,000	135
• Number of persons	70 persons @ £24,400	£1,708,000	124
• GFA (generally)	1,330 m² @ £1,207	£1,605,310	117
• GFA (non-air-conditioned)	1,330 m² @ £1,038	£1,380,540	101
• GFA (air-conditioned)	1,330 m² @ £1,374	£1,827,420	133
• GFA (1–2 Storey)	1,330 m² @ £1,033	£1,373,890	100

Notes: 1. GFA is based on 75% efficiency (1,000 NLA/0.75) = 1,330 m²
   2. Rates from the BCIS exclude external works and contingencies.

The range of budget calculations based on the various sources of costs is wide. The highest cost is 35% higher than the lowest budget calculation. Important considerations in reporting back to the design team and the client are the need to determine the likely efficiency of the building, which has a significant impact on the gross floor calculation; 75% efficiency has been assumed. If this increased to 85% then the GFA would drop to around 1,180 m² immediately having an effect on the overall GFA budget calculations. Similarly, if the efficiency were lower at 65% then a GFA of 1,540 would have to be factored into the calculations.

At this early stage and with the absence of decisions in certain key areas (number of storeys, building efficiency, air-conditioning or not) we shall be conservative in our recommendations to the design team and client. We shall assume for the purposes of our budget that the building will be air-conditioned, will not exceed 2 storeys and have a building efficiency of 75% as shown above.

Thus, our recommended building cost is based on a GFA of $1,330\,m^2$ at a rate of £1,374, which is the mean rate for air-conditioned offices. This rate also subsumes the 1–2 storey new build offices. So, our recommended budget for the building is rounded up to £1,900,000 based on a date of March 2006.

Before we complete the budget calculations there are a few issues we need to consider to enable us to arrive at a final budget figure. These issues centre upon building price indices and allowances for works outside the building that were not included in the rate per square metre adopted in the calculation above.

Differences in Dates

We must allow for the difference in general market price level between the date of tender of the building analysed/published and the date of preparation of this first estimate. Any difference in general market price level is allowed for by using an appropriate index. See Chapter 5 for a review of the type, use and discussion of the BCIS indices.

Question 6.7
Adjusting for Time Using Indices

Calculate the difference in building costs between the two dates below and the index numbers taken from the *BCIS Online* indices as shown:

(a) Cost index
BCIS general building cost index (excluding mechanical and electrical)
Base 1985 mean = 100

- January 1996: 166.1

- November 2005: 245.6

(b) TPI
BCIS all-in TPI
Base 1985 mean = 100

- 3rd quarter 1999: 151

- 4th quarter 2004: 223

Note that updating costs will always involve multiplying by the factor resulting from:

$$\frac{\text{The index at the later date}}{\text{The index at the earlier date}} = \text{Adjusting Factor}$$

If a percentage is required then the same formula applies as follows:

$$\frac{\text{The index at the later date} - \text{the index at the earlier date}}{\text{The index at the earlier date}} \times 100 = \% \text{ change}$$

When all the major differences between the analysed and the proposed building have been allowed for, an allowance should be made for the increase in general market price level which is expected between Strategic Briefing stage and the expected receipt of the tender for the new building. This should be carried out on the basis of the expert knowledge of market conditions of the quantity surveyor or cost planner.

One of the oldest indices in Australia is that in the *Building Economist Journal*. The AIQS has maintained this index for over 30 years. It is a national index, updated every quarter by the journal.

It is useful to quote the commentary from the *Building Economist* cost index as it provides important information on the limitations and use of indices in general.

> Statistical indices can do no more than measure national trends and, even though the general picture provides a useful starting point for dealing with specific projects, the particular circumstances of each scheme, its complexity or simplicity, local State conditions, tendering conditions and so on must always be considered when considering questions of building cost.
>
> This index (*Building Economist*) reflects the cost of employing labour and purchasing materials. It does not reflect tender values. Tender values fluctuate with market conditions and are influenced by main contractors' and subcontractors' profit margins, risk, shortage or oversupply of work, etc.
>
> The index should not be used for housing for which separate and special conditions would apply.

Preparation of the Budget Estimate

After examination of the client's initial brief, an inspection of the site and local conditions and an agreement of quality standards between the design team and the client, we must prepare our Strategic Brief stage budget estimate so that the client can decide whether or not to proceed with the project. As noted many times the decisions made during the early stages of a project are important and in this case we must decide whether to proceed and define what the budget or cost limit should be. It should be noted that the preparation of the estimate in this stage often forms part of a broader economic development appraisal. It may also have to explore the development alternatives that may be possible under the town planning scheme.

It is also possible that the client or the team involved with the economic feasibility stage have determined the budget (see budget setting using the development appraisal method) and this will form the cost framework for the project. When the budget has been established in this way, then our work will be similar to that described earlier. The major difference is that our focus will be on checking whether the functional costs and cost per square metre for our proposed project will be sufficient to allow the quality and scope of the works as the client and the design team anticipates.

Budget Calculations and Summary

New Offices: Southampton Present Date (March 2006) Prices		Total Cost £
1. New build offices Rate as discussed earlier taken from BCIS. Total area based on 75% efficiency as calculated above	1,330 m²@£1,374	1,827,420
2. Car parking No indication of type and extent of parking. Allowance based on past experience and similar projects. The number of spaces may be given in the brief or guidance may be sought from the local authority or planning scheme as to the required number of spaces for an office block of this size. Cost per space available in the BCIS.	Say	75,000
3. External works Assumed landscaping treatment/ crossover/drainage. Should be based on past similar projects and when sufficient information is available approximate quantities will be measured.	Allowance	100,000
4. External services Minimum requirement as services available in adjacent streets.	Allowance	50,000
		£ 2,052,420

5. Contract contingencies	+5%	102,620
At this stage of the project a reasonable allowance is made that takes account of the possible unforeseen work needed to complete the project		
		£ 2,155,040
6. Design risk	+5%	107,750
At this early stage a relatively large allowance is made to allow for uncertainty in the design and the design team.		
		£ 2,262,790
7. Updates		
(a) Present date, March 2006 to projected tender date of October 2006: Index 228 (forecast) to 234 (Forecast)	+2.6%	58,830
		£ 2,321,620
(b) Allowance for fixed price contract for an estimated 12 month contract (October 2006 = 234 to October 2007 = 247)	(Say) +5.6%	130,010
		£ 2,451,630

Notes: 1. Costs are rounded off to the nearest £10.
2. The BCIS index used is the all-in Tender Price Index (TPI).

Costs are rounded off to £2,450,000. A budget of £2,450,000 is considered reasonable, taking account of the information available.

Thus, the costs to be reported to the client for budgeting purposes are:

Total cost	£2,450,000	Cost per square metre (GFA = 1,330 m²) £1,842.11

Final Comments

Such an approach is open to criticism and opinions may vary as to the suitability of sources of information and rates chosen. However, we believe it is useful for you to follow the preparation of such an estimate at this stage of the design process using the BCIS as the primary source of cost information. We hope that the notes and commentary provided will make you appreciate the skill and judgement needed to prepare such an estimate yourself. It is also important that you prepare alternative estimates from different sources and judge what is the best to use in the prevailing conditions, location and for that type of building.

Do not see this example as a prescription for conducting all future estimates of this type. Individual project conditions must always be considered carefully and the needs of a specific project will always override a standard approach. Do not give unrealistic and possibly over-optimistic figures to design teams and clients just to encourage them to begin the project. Such an optimistic approach is a recipe for disaster!

So use this example with care and remember that unlike a real project much of the data was provided for you from the BCIS. In a real project you, or others, may have to spend a considerable amount of time collecting the basic data, discussing it with the design team, interviewing the client and other bodies such as the local authority. This information was provided conveniently for you in this example. In fact, collecting accurate data will probably take you a great deal longer than carrying out the estimating calculations given here.

Chapter Review

Stage A, the Strategic Briefing stage provides the starting point for all cost planning, because it is at this stage that the budget is set or the cost limit for the project is established. It is of crucial importance that the project is properly and fully evaluated using all the information that can be gathered and available at this early stage. The client must not be invited or misled into making a commitment beyond his or her financial capacity and it is incumbent on the design team to use their best efforts to arrive at a cost recommendation that is realistic and forms a sound foundation on which future cost planning can build. Clients who are encouraged to proceed with a project, which has been underpriced at this critical early stage are not likely to be impressed by the design team's excuses for cost overruns later in the design or tender stage. Sound professional judgement must be exercised and if this does not give the answer the client expected, then it should not be changed for expediency's sake.

The cost planner must relate the project information available to cost data that are obtainable in the office or in published sources. In this text we are going to make frequent use of the BCIS and some published sources of information such as *Spons, Laxtons, Rawlinsons, Cordell's* and *The Building Economist*. The important thing to remember is that the cost data must match the quality of the project information. The form of the cost data in the BCIS assists in this process of selection and elimination. The project characteristics have a major influence on the selection of an appropriate source and are determined primarily by the type or use of the building. Then within that building type follows the choice of a suitable quality with an identification of major difference between the projects. The final allowances are based on indices and forecasts of macro level effects on the project.

When the first estimate has been completed, it is put to the client as part of the design team's report on the feasibility of the project as a whole. If the client accepts this first estimate there will probably be lengthy discussion possibly entailing several modifications before acceptance. This is then regarded as the cost limit for the project from this stage onwards.

Although the first estimate during the Strategic Brief stage for most projects can be given as a lump sum, there will be other projects where the brief is not sufficiently developed or the technical difficulties are so great that a firm estimate cannot be given until after the completion of further work in the next stage, *Outline Proposals*. In these cases, the design team may well decide that the best approach at this early stage is to provide the client with a cost range within which the client is comfortable and the design team is confident of producing a design. The firm estimate will then be produced later when more information is available.

Review Question
Alternative Estimates

It is a useful exercise to re-price any estimate you prepare using a different source of cost information. This highlights the different cost base between the sources and identifies the similarity or the differences in the final estimate. If the differences are too large then the cost planner should find out the reason for the differences – is it the data or the calculations?

Re-price the functional unit superficial estimates for the new build offices given in this chapter using any on of the following sources in the UK:

- Langdon, D. (editor), (2006) *Spon's Architect's and Builders' Price Book*, E and F N Spon, London, UK.
- Johnson, V. B. (editor) (2006) *Laxtons Building Price Book*, Elsevier Press, Oxford, UK.
- Internal sources in the company or organisation for whom you work (if relevant) or from whom you can gain construction cost information.
- Any other cost publications you can gain access to, including possibly one of the large firms of quantity surveyors/construction cost consultants.

In Australia, use any of the following:

- *Rawlinsons Australian Construction Handbook.*
- *Building Economist.*
- *Cordell's Cost Guide.*

As a word of advice, read the preamble section of the source of your cost information. This is important as it provides details of the basis on which the costs are calculated and presented.

References

Akao, Y. (1990) An introduction to quality function deployment, in *Quality Function Deployment (QFD): Integrating Customer Requirements into Product Design*, Akao, Y. (editor), Productivity Press, Portland, Oregon, USA, pp. 1–24.

Atkin, B. and Flanagan, R. (1995) *Improving Value for Money in Construction: Guidance for Chartered Surveyors and their Clients*, Royal Institution of Chartered Surveyors, London, UK.

Austin, S., Baldwin, A., Li, B. and Waskett, P. (2000) Analytical Design Planning Technique (*ADePT*): a dependency structure matrix tool to schedule the building design process, *Construction Management and Economics*, **18** (2), pp. 173–182.

Australian Institute of Project Management/Construction Industry Development Agency (AIPM/CIDA) (1995) *Construction Industry Project Management Guide for Project Management Principals (Sponsors/Clients/ Owners), Project Managers, Designers and Constructors*, Australian Institute of Project Management, Canberra, Australia.

Australian Institute of Quantity Surveyors (AIQS) (2000) *Australian Cost Management Manual: Volume 1*, Australian Institute of Quantity Surveyors, Canberra, Australia.

Barrett, P. and Stanley, C. (1999) *Better Construction Briefing*, Blackwell Science, Oxford, UK.

Bennett, J. (1985) *Construction Project Management*, Butterworths, London, UK.

Bilello, J. (1993) *Deciding to Build: University Organisation and the Design of Academic Buildings*, Unpublished PhD Dissertation, Faculty of the Graduate School, The University of Maryland, USA.

Bolman, L. and Deal, T. (1997) *Reframing Organisations: Artistry, Choice and Leadership*, Jossey-Bass Publishers, San Francisco, CA, USA.

Bon, R. (1989) *Building as an Economic Process*, Prentice Hall, Englewood Cliffs, New Jersey, USA, p. 63.

Bresnen, M. J., Haslam, C. O., Beardsworth, A. D., Bryman, A. E. and Keil, E. T (1990) *Performance On Site and the Building Client*, Occasional Paper No. 42, The Chartered Institute of Building, Ascot, UK.

British Property Federation (BPF) (1983) *Manual of the BPF System*, British Property Federation, London, UK.

Building Economist (Quarterly Publication), *Journal of the Australian Institute of Quantity Surveyors*, Canberra, Australia.

Commonwealth of Australia (1978) *Comparative Capital Costs of Government and Non-Government Schools in Australia*, Australian Government Printing Service, Canberra, Australia.

Cornick, T. (1990) *Quality Management for Building Design*, Butterworths Heinemann, Guildford, UK.

Construction Industry Board (CIB, 1997) *Briefing the Team – A Guide to Better Briefing for Clients*, Thomas Telford, London, UK.

Construction Industry Development Agency (1993) *Construction Industry Project Initiation Guide for Project Sponsors, Clients and Owners*, Commonwealth of Australia, Canberra, Australia.

CSM (*Chartered Surveyor Monthly*) (1998a) **7** (9), p. 5.

CSM (*Chartered Surveyor Monthly*) (1998b) **7** (9), pp. 55–65.

Dell'Isola, A. J. (1982) *Value Engineering for the Construction Industry*, 3rd edition, Van Nostrand, Reinhold, New York.

Department of Industry (1982) *The United Kingdom Construction Industry*, HMSO, London, UK.

Department of Industry, Science and Resources (DISR) (1999) *Building for Growth: An Analysis of the Australian Building and Construction Industries* (*Competitive Australia*), Commonwealth of Australia, Canberra, Australia.

Dickey, J. W. (1995) *Cyberquest: Conceptual Background and Experiences*, Ablex, Norwood, New Jersey, USA.

Duckworth, S. (1993) Realising strategic decisions, *Journal of Real Estate Research*, **8** (Fall), pp. 495–509.

Egan, J. (1998) *Rethinking Construction*, Construction Task Force Report, Department of the Environment, Transport and Regions, HMSO, London, UK.

Ferry D. J., Brandon P. S. and Ferry J. D. (1999) *Cost Planning of Buildings*, 7th edition, Blackwell Science, Oxford, UK.

Friend, J. K. and Hickling, A. (1987) *Planning Under Pressure: The Strategic*, Pergamon, Oxford, UK.

Goodacre, P. E., Noble, B. M., Murray, J. and Pain, J. (1981) *A Design/Cost Theory for Measuring Buildings*, Occasional Paper No. 3, Design Evaluation, Department of Construction Management, University of Reading, UK.

Goodacre, P. E., Noble, B. M., Murray, J. and Pain, J. (1982a) *A Client Guide*, Occasional Paper No. 1, Department of Construction Management, University of Reading, UK.

Goodacre, P. E., Noble, B. M., Murray, J. and Pain, J. (1982b) *Client Aid Program*, Occasional Paper No. 5, Department of Construction Management, University of Reading, UK.

Gray, C. and Hughes, W. P. (2001) *Building Design Management*, Arnold, London, UK.

Gray, C., Hughes, W. and Bennett, J. (1994) *The Successful Management of Design: A Handbook of Building Design Management*, Centre for Strategic Studies, University of Reading, UK.

HMSO (1944), *Report of the Management and Planning of Contracts (The Simon Report)*, HMSO, London, UK.

Jaggar, D., Ross, A., Smith, J. and Love, P. (2002) *Building Design Cost Management*, Blackwell Publishing, Oxford, UK.

Johnson, V. B. (editor) (2006) *Laxtons Building Price Book*, Elsevier Press, Oxford, UK.

Kamara, J. M., Anumba, C. J. and Evbuomwan, N. F. O. (1999) Client requirements processing in construction: a new approach using QFD, *Journal of Architectural Engineering*, **5** (1), pp. 8–15.

Kamara, J. M. and Anumba, C. J. (2001) A critical appraisal of the briefing process in construction, *Journal of Construction Research*, **2** (1), pp. 13–24.

Keel, D. and Douglas, I. (1994) *Client's Value Systems: A Scoping Study*, The Royal Institution of Chartered Surveyors, London, UK.

Kelly, J., MacPherson, S. and Male, S. (1992) *The Briefing Process: A Review and Critique*, RICS Research Series, Paper No. 12, The Royal Institution of Chartered Surveyors, London, UK.

Kelly, J. and Male, S. (1993) *Value Management in Design and Construction*, E and F N Spon, London, UK.

Latham, M. (1994) *Constructing the Team: Joint Review of Procurement and Contractual Arrangements in the United Kingdom Construction Industry*, Final Report, HMSO, UK.

Langdon, D. (editor), (2006) *Spon's Architect's and Builders' Price Book*, E and F N Spon, London, UK.

Mann, T. (1992) Chapter 3 – Initial building (construction) cost, in *Building Economics for Architects*, Van Nostrand Reinhold, New York, pp. 15–26.

Masterman, J. W. E. (2002) *An Introduction to Procurement Systems*, E and F N Spon, London, UK.

Ministry of Public Buildings and Works (1968) *Cost Control During Building Design*, HMSO, London, UK.

Ministry of Public Buildings and Works, Research and Development (1970) *The Building Process: A Case Study from Marks and Spencer Limited*, R&D Bulletin, HMSO, UK.

Morton, R. (2001) *Construction UK: Introduction to the Industry*, Blackwell Science, Oxford, UK.

Mohsini, R. A. (1996) Strategic design: front end incubation of buildings, in *North Meets South: Developing Ideas*, Taylor, R. G. (editor), Proceedings of CIB W92 – Procurement Systems, Department of Property Development and Construction Economics, University of Natal, Durban, South Africa, pp. 382–396.

Murray, J. P., Gameson, R. N. and Hudson, J. (1993) Creating decision-supporting systems, in *Professional Practice in Facility Programming*, Preiser, W. F. E. (editor), Van Nostrand Reinhold, New York, pp. 427–452.

Murray, M. and Langford, D. (2003) *Construction Reports 1944–98*, Blackwell Science, Oxford, UK.

National Economic Development Office (1983) *Faster Building for Industry*, HMSO, London, UK.

National Economic Development Office (1983) *The Professions in the Construction Industries*, HMSO, London, UK.

Newman, R., Jenks, M., Bacon, V. and Dawson, S. (1981) *Brief Formulation and the Design of Buildings (Summary Report)*, Department of Architecture, Oxford Polytechnic, UK.

Newton, S. (1989) *Cost Management and the Project Manager*, Unpublished Paper, University of Technology, Sydney.

NPWC (1989) *Guidelines for Cost Planning Consultant Services*, Australian Government Publishing Service, Canberra, Australia.

O'Reilly, J. N. (1987) *Better Briefing Means Better Buildings*, Building Research Establishment, Garston, UK.

Property Council of Australia (1997) *Method of Measurement*, Property Counsil of Australia, Sydney, Australia.

The Property Services Agency (1981) *Cost Planning and Computers*, Department of the Environment, London, UK.

Rawlinsons (Annual Publication), *Rawlinsons Australian Construction Handbook*, Rawlhouse Publishing, Perth, Western Australia, Australia.

Royal Institution of British Architects (1998) *Architect's Handbook of Practice Management*, 6th edition, RIBA Publications, London, UK.

Royal Institution of British Architects (RIBA) (2000) *The Architect's Plan of Work*, RIBA Publications, London, UK.

Salisbury, F. (1998) *Architect's Handbook for Client Briefing*, Butterworth, London, UK.

Saaty, T. L. (1990a) *Multi Criteria Decision Making: The Analytic Hierarchy Process*, Vol. 1, AHP Series, RWS Publications, Pittsburgh, Philadelphia, USA.

Saaty, T. L. (1990b) *Decision Making for Leaders*, Vol. 2, AHP Series, RWS Publications, Pittsburgh, Philadelphia, USA.

Seeley, I. H. (1996) *Building Economics*, 4th edition, Macmillan, London, UK.

Smith, J. (2005) An Approach to Developing a Performance Brief at the Project Inception Stage, *Architectural Engineering and Design Management*, **1** (1), pp. 3–20.

Smith, J. and Love, P. (2000) *Building Cost Planning in Action*, University of New South Wales Press, Sydney, Australia.

Thiry, M (1997) *A Framework for Value Management Practice*, Project Management Institute, Washington, DC, USA.

Thompson, A. A. and Strickland, A. J. (1995) *Strategic Management: Concept and Cases*, 8th edition, Irwin, Chicago, Illinois, USA.

United States General Accounting Office (1978) *Computer-Aided Building Design*, US Department of Commerce, Washington, DC, USA.

Walker, A. (2002) *Project Management in Construction*, 3rd edition, Blackwell Science, Oxford.

Woodhead, R. M. (2000) Investigation of the early stages of project formulation, *Facilities*, **18** (13/14), pp. 524–534.

Woodhead, R. (1999) *The Influence of Paradigms and Perspectives on the Decision to Build Undertaken by Large Experienced Clients of the UK Construction Industry*, Unpublished PhD Thesis, School of Civil Engineering, University of Leeds, UK.

Woodhead, R. and Smith, J. (2002) The decision to build and the organization, *Structural Survey*, **20** (5), pp. 189–198, Emerald Press, Bradford, UK.

Wyatt, R. (1999) *Computer-Aided Policy Making: Lessons from Strategic Planning Software*, E and F N Spon, London, UK.

Yang, J. and Lee, H. (1997) An analytic hierarchy process decision model for facility location selection, *Facilities*, **15** (9/10), pp. 241–254.

Further Reading

AIQS (2000) *Cost Control Manual*, National Public Works Council Inc., Canberra.
    Sub-Section 5.3: Building price indices.

Ashworth, A. (2004) *Cost Studies of Buildings*, edition, Longman Scientific and Technical, Harlow, England.
    Chapter 5: Pre-tender price estimating, pp. 77–101.

Building Economist, any volume, refer to the Cost Index, The Australian Institute of Quantity Surveyors, Canberra.

Commonwealth of Australia (1978) *Comparative Costs of Government and Non Government Schools in Australia*, Australian Government Publishing Service, Canberra.

Ferry, D. J., Brandon, P. S. and Ferry, J. D. (1999) *Cost Planning of Buildings*, 7th edition, BSP Professional Books, Oxford.
    Chapter 11: Introduction to Cost Modelling, pp. 109–119.
    Chapter 14: Cost indices, pp. 176–196.
    Chapter 15: Cost planning the brief, pp. 197–212.

Flanagan, R. and Tate, B. (1997) *Cost Control in Building Design*, Blackwell Science, Oxford.
    Chapter 4: Estimating at the early design stage, pp. 43–48.
    Chapter 5: Cost control during inception, feasibility and outline proposals, pp. 46–56.
    Chapter 10: Cost plannning during feasibility, outline proposals and scheme design, pp. 151–156.
    Chapter 11: Cost example from feasibility to scheme design, pp. 227–242.

Morton, R. and Jaggar, D. (1995) *Design and the Economics of Building*, E & F N Spon, London, UK.
    Chapter 13: Cost prediction – science or guesswork? pp. 289–314.

SUGGESTED ANSWERS

Question 6.1: Functional Unit Method of Estimating

(a) Hotel

95 Bedrooms × £57,000 per unit = £5,415,000

(b) Secondary school

450 Pupils × £7,430 per pupil = £3,343,500

These functional unit estimates are dangerously simple to prepare! The skill is not in executing the calculation, but in choosing a suitable unit rate for your particular project. The above estimates are not complete, nor fully inclusive. You must read carefully the information given in the preamble to the section of the cost source you are using. Discover what the cost per functional includes and excludes.

In the case of the *Building Economist* (Australia), it warns that the 'costs given are average prices for typical buildings within metropolitan areas.' The prices given also exclude land costs, external works, external services more than 3.0 m from the outside face of the building, landscaping, special equipment, legal and professional fees, contingencies, locality allowance(s), rise and fall (fluctuations) allowances during the contract period and inflationary allowances (if any) from the date of the preparation of the estimate to the projected date of the tender. Suitable provision for all or some of the items noted here may result in an additional and substantial amount of 25–50% being added to the above 'net' estimate.

Question 6.2: Superficial Budgets

(a) Hotel

5,200 m² GFA × £1,120/m² = £5,824,000

(b) Secondary school

3,600 m² GFA × £1,060 per pupil = £3,816,000

The GFAs may be derived from analyses of previous similar schemes, from town planning criteria or organisational standards and in-house analyses of the target areas per functional unit. So, for instance, the area per pupil in schools may be analysed and an average of mean area per pupil derived to provide future planning guidelines for area standards.

As earlier, the total cost or budgets calculated are exclusive of external works and services, land, contingencies, furniture and fittings and many other similar items that must be separately costed.

Question 6.3: Offices in Liverpool: Superficial Method of Estimating

- Net Rentable Area (NRA)   1,260 m^2

- Add   Allowance for circulation space, lift lobbies,
           public areas, plant rooms, etc. 20% of GFA.
           (Based on analyses of similar buildings) = ÷4   315

  Total GFA   1,575 m^2

- Cost per square metre GFA (Liverpool)   £1,200

  Add 10% for office buildings with higher standards   £120

  Total cost per square metre   £1,320

  Total cost of building = 1,575 m^2 @ £ 1,320 =   £2,079,000

The same comments given at the end of Question 6.1 apply equally to this estimate.

Question 6.4: Offices in Slough: Superficial Method of Estimating

Inspection of the BCIS extract in the text (Table 6.10) shows that 'new build air-conditioned offices' in March 2006 for a UK mean location shows a median rate per square metre of £ 1,287.00 and a mean rate of £ 1,374.00. The range of rates in this category is from £ 518.00 to £ 3,504.00 and based on a healthy sample of 281 projects.

Now we have to decide which rate to use based on the information we have gained from the BCIS data, or any other source. This process is not straightforward as we have little design information available. Our project is unlikely to be at the extremes of the costs range and so we are going to play it safe and use a rate slightly above, but close to the mean rate of £ 1,374.00. We are going to use a rate of £ 1,400.00/m^2 GFA Building cost per square metre, taken from the BCIS or from either *Spons, Laxtons* in the UK, or *Rawlinsons, Cordell's* or the *Building Economist* in Australia. You may use any rate that you consider appropriate for this scheme.

Total Building Costs 1,780 m^2 × £ 1,400.00 =   £ 2,492,00

Add   External works/car parking (separate estimate)   £ 252,250

         Total Costs   £ 2,744,250

Again, this estimate is still not complete and the notes in Question 6.1 should be heeded. Ideally, use more than one source to check the accuracy of your budget setting.

Question 6.5: Functional Area Estimate

The estimate is prepared in a similar format to that of the source information. The next step is to estimate the cost of each functional cost based on the present date costs given. The areas of the circulation spaces associated with each group area were all given as nil in the question. Therefore, these areas are not priced in the following estimate. The areas for travel/engineering will take care of the total circulation space for the school.

Functional spaces	Cost $£/m^2$	Area	Total cost $£$
1. Library Resource Centre			
Main Reading Area	696	55	38,280
General Workroom/Store	858	45	38,610
Circulation	nil	–	–
2. Learning Areas			
Classrooms	672	386	259,392
Practical Activities	875	465	406,875
Withdrawal Room	823	74	60,902
Circulation	nil	–	–
3. Administration/Staff			
Principal's Office	683	18	12,294
Assistant Principal Office	666	10	6,660
Reception/Waiting Area	603	12	7,236
Store Rooms	777	13	10,101
Staff Lounge	653	36	23,508
Staff Toilets	1,362	12	16,344
Student Rest Room/Toilet	1,203	10	12,030
Cleaner's Store	1,401	4	5,604
Circulation	nil	–	–
4. Pupil's Amenities			
Pupil's Toilets	1,230	52	63,960
Circulation	nil	–	–
5. Travel/Engineering	196	410	80,360
Total Building Costs			£ 1,042,156

Note that these are the net building costs only based on the present date costs. The costs per square metre listed here can be used as a basis for any revisions to the scheme that result in changes to the floor area of each functional space.

The format of this estimate lends itself to the transfer of information to a spreadsheet, which would speed up the estimating process and enable the cost planner to accommodate changes in the brief and layout and cater for changes much more rapidly.

Question 6.6: Methods of Estimating for Cost Planning and the RIBA Stages

The methods are described in the chapter and the further reading, but the essence of each method is given below.

As for the RIBA stage at which each method is used the important factor to remember is that a technique can only be used when suitable information to operate that technique becomes available. Similarly, when more detailed project data is produced by the design team the cost planner should not continue using a coarse method of estimating, which has been clearly superseded in application by a more accurate technique that uses the more detailed information in an appropriate way.

Unit method (budget setting)
This is a single rate method of estimating based on the cost of a suitable functional cost per place. The method is based on the fact that a close relationship exists between the cost of a building of a particular type and the number of functional units it accommodates. Cost data is collected from previously completed similar projects and the total costs are divided by the number of functional units provided in the project. The total cost per unit is then applied to similar cost planned projects, with appropriate adjustments for time from the original. Quality adjustments are difficult to make with such a coarse method of estimating. As in any single rate method of estimating the choice of a unit rate is of critical importance; small errors in the rate can cause a substantial variation in the prediction of the total costs.

Stage used: An early stage estimating method
RIBA Stage A Options Appraisal
RIBA Stage B Strategic Briefing
See Figure 6.3.

Superficial area method (Budget Setting)
This is another single rate method of estimating subject to the prediction problems noted above. Costs per square metre of the GFA are readily

obtained from published sources (BCIS and price books) and easily calculated from completed projects. The superficial area method is a popular technique because it correlates cost to an easily recognised quantity. The floor area may be measured from early drawings, or it may be derived from the client's statement of requirements. This is particularly the case with commercial clients in office and retail space, where they often have an early and clear idea of their requirements in terms of floor space. Accuracy of this method is similar to the *unit method* and again, is subject to a great deal of sensitivity to the actual rate per square metre chosen. The rate selected must be adjusted to take account of time, market conditions, risk and other local factors.

Stage used: Another early stage estimating technique.
              RIBA Stage B Strategic Briefing.
              There is also a possibility that if the design work and other project decisions are not well advanced that this technique may also continue to be used in to the next Stage C – Outline Proposals. However, this is not to be encouraged, as more data suitable for a more rigorous technique should be available by this stage.
              See Figure 6.3.

Functional area method (Budget Setting)
This method is a development and a refinement of the single rate of estimating used in the above techniques. It reflects design activities and provides a bridge between the early-stage methods and the more detailed elemental estimating methods described later. This method should be more accurate than the single rate methods. It is only suitable for projects of a similar type where the investment of time in analysis to gain the cost per functional area can be used on a number of future projects in relatively large building programmes such as schools or hospitals.

The costs per square metre of functional space for the type of project are applied to the areas defined in the proposed project. The areas in the new project may be measured, if sketch plans have been prepared, or if no measurable information exists then the areas are gained from a client statement of functional areas contained in the brief.

Stage used: Another early-stage estimating technique that can be used in two of the RIBA stages:

              • Stage B Strategic Briefing

              • Stage C Outline Proposals

                See Figure 6.3.

Question 6.7: Adjusting for Indices

Calculate the difference in building costs between the two dates below with the index numbers taken from the *BCIS Online* indices as shown:

(a) Cost index
BCIS general building cost index (excluding mechanical and electrical)
Base 1985 mean = 100

- January 1996:   166.1

- November 2005: 245.6

$$\text{Percentage change} = \frac{245.6 - 166.1}{166.1} \times 100 = \frac{79.5}{166.1} \times 100 = 47.86\%$$

$$\text{Factor change} = \frac{245.6}{166.1} = 1.479$$

(b) TPI

BCIS all-in TPI
Base 1985 mean = 100

- 2nd quarter 2001: 170

- 4th quarter 2004: 223

$$\text{Percentage change} = \frac{223 - 170}{170} \times 100 = \frac{53}{170} \times 100 = 31.18\%$$

$$\text{Factor change or weighting} = \frac{223}{170} = 1.312$$

Review Question

Alternative estimates
Prepare an alternative estimate using the various sources of published cost data. In the UK, use *Spons* and *Laxtons*. In Australia, use *Rawlinsons* and *Cordell's*.

The purpose of this question is for you to gain access to the alternative sources of cost information, and more importantly, be aware of the basis on which these costs are given.

The questions you must answer are:

- Does the estimated costs vary between each publication for the same type of building?

- If it does vary, is it significant and what are the reasons for the differences?

- Are the alternative sources simple to understand and to apply in practice?

- Is the range of building types adequate and comprehensive for future application?

- Does the second source provide advice on general and historical cost movements through the use of a cost index?

The important thing is that you must be aware of alternative sources of cost information and you should be flexible in your approach. Whilst you may have a preference for one particular source (and we all tend to develop preferences for certain sources) you should always carry out a check on your estimate using one or more of the alternatives. Following this practice may identify and isolate a deficiency that your single source did not recognise. The revised costing using an alternative should be carried out quickly once you have the basic data and provides a valuable check on your costing. The use of a wide a range of cost sources should be encouraged and practised rigorously throughout your cost planning career.

Chapter 7

Cost Planning the Outline Proposals Stage

Running through the whole process of cost planning is the search for the optimum solution that will offer the best value for money or lead to economy and savings ... Cost planning is, therefore, a consistent and systematic framework for estimating; without it there is no scientific basis of comparative analysis and whilst estimates are at risk until proved, there is less chance of error when using such a system.

Junior Organisation Quantity Surveyors Standing Committee (1976: 2)

Chapter Preview

The outline of the approach used in this chapter and the following three chapters is broadly summarised in Table 7.1. This table used in Chapter 7 provides the guide to all the Royal Institute of British Architects (RIBA) stages discussed in Chapters 7 to 9 and it will be used to identify the specific stage being described. The focus of this chapter is the *Outline Proposals* stage and this shown as the shaded portion in Table 7.1.

As noted in the previous chapter the descriptions of each stage will emphasise the theory of cost planning and will provide an example of cost planning during each of these stages. In this way the theory can be related to the practice. For simplicity and ease of understanding the stages outlined here are shown in a linear sequence. However, in practice this linear

Table 7.1 RIBA Outline Plan of Work 1998

Pre-design (briefing)			Design			
Inception or Feasibility			Pre-construction Period			
Pre-stage A	Work Stage A	Work Stage B	Work Stage C	Work Stage D	Work Stage E	Work Stage F
	1	2	3	4	5	6
Establishing the Need	Options Appraisal	Strategic Briefing	Outline Proposals	Detailed Proposals	Final Proposals	Production Information
	Chapter 6		Chapter 7	Chapter 8	Chapter 9	
	Briefing		Outline Proposals	Detailed Proposals	Documentation	

sequence can be more unstable and non-sequential to reflect new or unexpected information becoming available to the design team.

This chapter starts the budget distributing process of the design stage first identified in Chapter 6. In that chapter the various techniques, processes and approaches to estimating and cost planning were broken down into two basic types:

1. Budget setting
2. Budget distributing.

The first set of techniques we have focused on up to now budget setting have been largely confined to the pre-design stages where the project parameters were being established. Once we have set the cost limit or budget at the Strategic Briefing stage the emphasis switches to taking this sum of money embodied in the cost limit and distributing it to the various parts of the project to gain a balanced distribution of costs and to achieve the best value for that defined sum of money.

This chapter begins with a review of the budget distributing techniques used in the design stages and these are summarised in Figure 7.1, first used in Chapter 6 to identify the use of the various Cost Planning techniques in the stages of the RIBA Plan of Work.

OUTLINE PROPOSALS: PLAN OF WORK STAGE C

The Outline Proposals stage C is the third formal stage in the RIBA Plan of Work and the first step in the design stage of the pre-construction cost management process. Inspection of Table 7.1 shows its position (shaded) in the design process. The Outline Proposals stage is a concentrated period of developing and assessing a range of possible alternatives to satisfy the brief.

It has to be recognised that the design and other data that will affect costs in a significant way will still be sparse unless the design team is committed to expediting each stage. In fact, in practice, for some projects cost planning proper may not commence in earnest until the next stage D, Detailed Proposals. We realise that some design teams may not be very keen to perform any cost planning during this stage of design because their overriding concern may be to focus on the development of alternative proposals. However, this is not to be encouraged and could prove dangerous if we are to succeed in maintaining our Cost Planning principles. To be effective, cost planning must be performed in steps of increasing detail that are integrated with the expanding information base and the increasing detail of the developing design decisions.

Pre-design			Design			
Inception or Feasibility			Pre-construction Period			
Pre-stage A	Work Stage A	Work Stage B	Work Stage C	Work Stage D	Work Stage E	Work Stage F
	1	2	3	4	5	6
Establishing the Need	Options Appraisal	Strategic Briefing	Outline Proposals	Detailed Proposals	Final Proposals	Production Information
EARLY STAGE TECHNIQUES		Cost Limit	Cost Targets		Cost Checking	
(Described later in Chapter)						
• Strategic Value Management						
• Strategic Needs Analysis						
• Quality Function Deployment						
• Situation Structuring						
	• Expert Choice					
	BUDGET SETTING TECHNIQUES		BUDGET DISTRIBUTING TECHNIQUES: BUDGET DISTRIBUTION, CHECKING AND RECONCILIATION			
	• Developer's Budget					
	• Total cost					
	• Functional Unit (student, bed, seat, room)					
		• Superficial (gross floor area)				
		• Functional Area/Cost				
			• Elemental (group)			
				• Elemental (individual and sub-elemental)		
			• Approximate Quantities (to suit level of detail available)			

Figure 7.1 Cost Planning Techniques and the Outline RIBA Plan of Work

In our concept map first introduced in Chapter 4 (Figure 4.2) the Outline Proposals stage is the confirmation of the budget or cost limit for the project and the start of our cost planning journey. Instead of leaving the cost limit as a lump sum we can begin to divide this global sum into 'packages' of group element cost. Although a modest beginning, this commences our cost planning of the project. This is illustrated in Figure 7.2 where the concept plan shows the position of this stage in relation to the subsequent stages.

OVERVIEW OF ACTIVITIES AND DOCUMENTATION

The RIBA (2000: 16, 17) Plan of Work identifies a series of overlapping cost management activities taking place at this stage. These can be summarised as:

- evaluate the strategic brief;

- assess the economic constraints;

- advise on project programme;

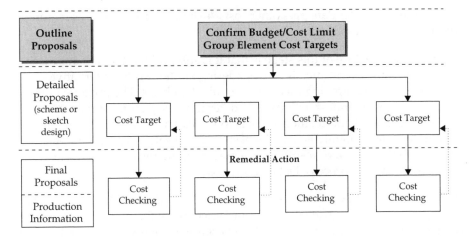

Figure 7.2 Cost Planning Concept Map: Outline Proposals Stage
Source: Adapted from Smith and Love (2000: 11).

- implement design and cost management procedures in programme;

- participate in the development of the project (design) brief;

- advise on the cost effect of design and energy options;

- prepare initial cost plan;

- prepare initial cash flow projection;

- advise on procurement options.

In addition, another important task at this stage is that the design team is also focused on all the work needed to submit and obtain detailed planning approval.

Alternative Outline Proposals will be developed and are considered at this stage. The information at this stage will consist of:

- the expanded strategic brief;

- sketches and other relevant information;

- further information may be available about user requirements, site conditions, planning requirements, design approach and other studies.

The end result of this stage is a cost plan presented in a group element format to provide the first attempt at allocating costs to major parts of the project. All costs presented in this cost plan are to be current project cost plus escalation to the forecast tender date plus other costs such as external works as required, each separately identified.

An important feature of this stage is that it is still probing the best approach to satisfying the client's needs by considering all the feasible alternatives considered and comparing them on an equivalent basis. The purpose of the design and cost planning activity is to determine the general approach in functional content and relationships, layout, design and construction. The design team must jointly isolate the best means of satisfying the strategic brief, with the chosen or resultant scheme becoming the basis for the next stage in the design process, Detailed Proposals.

Good communication is essential to this whole process and the client must be closely involved in the development and selection of the final scheme. At the end of this stage client approval of the chosen scheme is essential for the next stage to commence.

The method of approach by the cost planner is to work closely with the design team in developing the alternative proposals and to establish comparative costs for the various schemes using the superficial method of estimating at commencement of the stage, but if the type of scheme permits, an approach using the functional area costing approach may be adopted. In this way the evolving brief and the different functional spaces in each scheme can be more accurately reflected in the final costs. The alternative Outline Proposals schemes can be costed in relation to size and type until the most effective solution is found. External works and other costs beyond the external walls of the building are shown separately.

THE APPROACH TO THE OUTLINE PROPOSALS STAGE C

The information base of the project continues to expand during this stage as more design team and client interaction takes place. The requirement to submit a planning application (if a permit is required) will prompt the necessity to make some important decisions on the building type (use and uses in the building), gross floor area (GFA), number of storeys, height of building, orientation of building, location of building on the site, setbacks, site coverage, access to and from adjoining streets and other issues that must be made when making a planning application. Where a planning permit is not required (a permitted use under the planning scheme) the cost planner should be encouraging the design team to address these basic design issues at the earliest date if they have not already done so.

ADDITIONAL INFORMATION AVAILABLE AT THE OUTLINE PROPOSALS STAGE

Further information is usually available at the beginning of Outline Proposals that was not available at the beginning of the previous stage, Strategic

Briefing. More information becomes available regarding the following issues in this stage:

- design criteria;

- site information;

- environmental and design constraints;

- planning requirements for outline development control approval (if required);

- energy targets;

- programme planning requirements including work stage procedures;

- the development of the project brief with all the additional information that this document requires.

For the cost planner on a commercial project such as offices, retail, industrial and hotels, one of the most important pieces of financial information that has been developed at the end of the Strategic Briefing stage is the Feasibility or Development Appraisal Report (including the indicative cost or maximum capital cost, which has been approved by the client by this time). The preparation of this report should be integrated into the Strategic Briefing stage as noted earlier and the results should form the financial and to a great degree the design constraints for this stage. The capital cost limit for the project should be confirmed in the development appraisal (feasibility study) and the type and scope of the project should be clearly established. The parameters established in the development appraisal should be clearly communicated to all design team members to provide a framework for their work in this stage. That is, the strategic issues should be firmly connected into this first design stage and not discarded or ignored.

As noted in Chapter 6 a detailed review of feasibility studies does not form part of this Cost Planning description and they are comprehensively covered in the literature (Robinson, 1989; Cadman and Topping 1995; Isaac, 1996, 1998; Brown and Matysiak, 2000). Smith and Love (2002: 15–21) also illustrate a development appraisal integrated into the Strategic Briefing stage.

Although there is an increase in the amount of design information there is still not enough data for reliable cost targets to be prepared for each individual element. So, for this stage an outline cost plan is prepared with the total cost limit broken down into group elements, rather than cost allocations made to all the individual elements.

GROUP ELEMENTS

A cost target is prepared for each group of elements. These group elements for the UK and Australia are listed in Table 7.2.

The allocation of costs to the group elements has the advantage of forcing the design team to think clearly at this early stage about the broad allocation of expenditure to the major parts of the building.

Before preparing an outline cost plan, an allowance should be made for unforeseen planning and design difficulties, which may come to light later in the design process and for price rises between the preparation of the outline cost plan and the submission of tenders. Note that the allowance made at this stage for these price rises could be different to the corresponding allowance included in the Strategic Brief stage because it is an allowance made at a later date and possibly with better information about the projected tender date.

Table 7.2 Group Elements for the UK and Australia

BCIS/RICS Group Elements (UK)	AIQS Group Elements (Australia)
1. Substructure	1. 00PR Preliminaries
2. Superstructure	2. 01SB Substructure
3. Finishes	3. 02CL–05RF Superstructure
4. Fittings and Furnishings	4. 06EW–08ED External Fabric and Finishes
5. Services	5. 09NW–11ND Internal Fabric
6. External Work	6. 12WF–14CF Internal Finishes
7. Preliminaries	7. 15FT–16SE Fittings
8. Contingencies	8. 17SF – 29SS Services
	9. 30CE Centralised Energy Systems
	10. 31AR Alterations
	11. 32XP–36XL Site Works
	12. 37XK–44XS External Services
	13. 45XX External Alterations
	14. 46YY Special Provisions

ALLOWANCES FOR CONTINGENCIES

Potentially during the cost planning stages from Outline Proposals onwards there are four contingency factors to be considered by the cost planner and the design team:

1. Planning contingency

2. Design (risk) contingency

3. Contract contingency

4. Project contingency.

The AIQS (2000: 2–17) has defined these contingencies as follows:

> The Planning Contingency is an allowance to cover the risk of not being able to design the spatial relationships and achieve the desired functional area and travel/engineering allowances. (Planning Contingency) … will be reduced to zero at Sketch Design (Detailed Proposals).

> The Design Contingency is an allowance to cover the risk of the estimator/cost planner not adequately foreseeing the correct design or the complexity of the design. The amount of the Design Contingency will depend on the amount of detail available, and will be reduced to zero at Tender Document stage (Production Information).

> The Contract Contingency is an allowance to cover the risk of variations and unforeseen items encountered *during construction.*

> A Project Contingency may also be added to cover delays and/or inflation, major changes required by the client or authorities, fee negotiations and similar.

In the UK, the Building Cost Information Service (BCIS) only recognise the contract contingencies as defined above. The remaining definitions are in the province of cost planning (not cost analyses) and are therefore at the discretion of the cost planner. In our cost planning approach we believe it is prudent to make the necessary allowances, or at least consider their inclusion, in the cost plan to the conclusion of the stages noted in the definitions.

Before preparing our outline cost plan we should decide the amount of our contingency or reserve against risk of changes in price (contract sum) or design under the four headings noted above.

The AIQS (2000) shows the Design and Contract Contingencies as a separate heading on the cost planning standard forms. The Project Contingency, if required, should be included as one of the headings in the Special Provisions. The Planning Contingency is an allowance reserved for the Outline Proposals stage and is assumed to be unnecessary from the Detailed Proposals onwards. These sums are best shown as separate sums in the cost plan and not included in the cost and in the rates for our elemental costs.

The percentages allocated to Planning, Design, Contract and Project Contingencies should be assessed, normally as a *global* percentage and included in the cost plan under their own separate heading, so that it is clear to all the design team and the client the assessment that has been made for each of these categories of contingency. The cost planner has to assess these allowances for each project taking account of the project environment and not merely use a standard percentage regardless of the project context. The percentage allowance should be assessed for the possible changes that can occur in the price of the project and for unforeseen problems both during all the design stages and for the contract contingency during the construction period. Normally, a well thought out scheme would carry a smaller percentage than one where many problems have to be solved by the design team. As the scheme progresses through the various stages of design development it can be expected that the percentage will fall, and in some cases disappear to reflect the greater certainty of the design decisions.

The Contract Contingency is the amount or allowance that carries forward into the tender. Like the other contingencies it generally reduces in magnitude as the design progresses, until at the Tender Stage it would reflect the percentage to be included in the contract documents. Thus, in normal circumstances the percentage may range from around say 5% in the early stages to fall to possibly as low as 1% or 2% in the Tender Document stages, subject to assessments of the possibility of unforeseen problems. Note that these are indicative figures and cannot be considered prescriptive and fixed amounts in every project. Circumstances may exist where the order of difficulty and complexity require a higher order of percentages to take account of the greater degree of risk involved. Similarly, on a repetitive design with a design team used to a particular design and construction system on a normal site the percentage is likely to be lower.

ESCALATION TO TENDER DATE

We must also consider how prices are expected to move between the preparation of this outline cost plan and the projected tender date. The construction project has to be assessed carefully using the opinions of any expert views in the general and professional press. The BCIS Building Price Indices should be consulted for past performance of costs and as a means of projecting any changes between the date of the cost plan and the future tender date. As noted in the previous chapter the BCIS indices provide a comprehensive source for past index values based on a specific date and they are also a significant source of forecast or predicted index values in the future. In fact, as noted in the last chapter functional costs, cost per square metre and whole cost analyses can be re-indexed to the date nominated by the cost planner to reflect the present date, tender date and completion date as required.

The cost planner should make provision for any changes in cost to the expected tender date in the summary of the cost plan under an appropriate heading.

In Australia, the preamble to the Current Construction Costs Guide in the *Building Economist* contains a quarterly review of market conditions in all the capital cities and the *Rawlinsons Australian Construction Handbook* has details of its own indices and projections in quarterly updates over the coming months. The escalation percentage from the date of the cost plan to the projected tender date is noted as *Escalation to Tender Date* on AIQS (2000: A1–6, 7) Cost Planning Forms 2 and 3.

Escalation During Construction Period

The last global adjustment we have to make is for the estimated allowance for fluctuation in the tender sum (*rise and fall* allowances in Australia) during the contract period. The cost planner should make provision for any changes in cost between the expected tender date and the completion date in the summary of the cost plan under an appropriate heading. Using the BCIS indices this change can be made using the projections of the indices to the end of the construction period. This makes this adjustment in the UK simple and easy to make as shown in the cost plan presented in Chapter 6.

In Australia, the adjustment has to be made 'manually' by the cost planner using judgement and experience to interpret and forecast the possible direction and magnitude of the indices in the future. This is the last adjustment to be made on the cost plan (see AIQS (2000: A1–6, 7) Cost Planning Forms 2 and 3). This allowance is the cost planner's assessment of the amount the successful contractor allows for increases (or decreases) in cost between the date of tender and the completion of the project. If the contract includes a fluctuations (*rise and fall*) clause then this amount will be paid during the progress of the works in interim certificates or progress payments. If the contract is a fixed price one, then the cost planner has to assess the amount the estimator will allow in the tender for such changes in price. Whether the contract allows for price changes or is fixed price, the cost planner has to predict the construction market from tender to completion of the project and make an allowance for expected price fluctuations during this period.

Whatever the method used to predict the final cost of the project the allowance or sum calculated made will be related to the length of the construction time for the project and the risk involved. Naturally, it will increase the longer the time needed to complete the works.

> **Question 7.1**
> **Outline Cost Plan Allowances**
>
> An outline design of a simple office building has a floor area of 2,200 m² GFA (as measured from the Outline Proposal drawings).
>
> The Brief Stage budget is £1,870,000.
>
> Make allowances as noted above for the following:
>
> - Planning contingency
> - Design contingency
> - Contract contingency
> - Escalation to tender date
> - Escalation during contract period.

GROUP AND ELEMENTAL COST TARGETS

We will now consider how cost targets are prepared to each of the group elements as shown earlier.

It is likely that one or more of several situations will exist where:

- a cost analysis of similar building available from the cost planner's own sources; or

- a cost analysis, or cost analyses, of similar buildings are available from a database such as the BCIS; or

- published data is the only available source.

Figure 7.3 summarises the activities taking place at the *Outline Proposals* stage leading to the preparation of the outline cost plan within the cost control principles noted earlier.

GROUP ELEMENT RATES USING BCIS SOURCES

The BCIS should be interrogated to find a range of analysed buildings similar to the project proposal. With the information that is available for the new project this should be used to identify projects with similar characteristics. The first step in this process is to define the use or function of the analysed and new buildings:

- function(s) or use(s),

- new build or renovations,

- area or space,

- number of storeys,

- location,

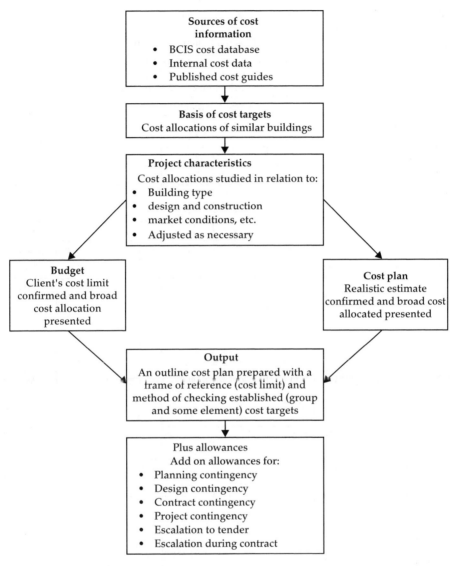

Figure 7.3 Summary of Outline Proposals Stage
Source: Adapted from Ministry of Public Buildings and Works (1968: 66).

- general statement of quality,

- type of construction.

A condensed version of the BCIS analysis of the rates per square metre for schools under the categories is shown in Table 7.3:

- generally (all schools irrespective of level)

- primary

- secondary.

Table 7.3 BCIS Group Element Prices per Square Metre

Group Element Cost per Square Metre Study							
Rate per square metre gross internal floor area for the group element excluding external works and contingencies and with preliminaries apportioned by cost. Last Updated 1 April 2006. At 4Q2005 prices (based on a Tender Price Index of 223) and UK mean location.							
Building Function	Mean	Lowest	Lower Quartile	Median	Upper Quartile	Highest	Sample
New Build							
710. Schools							
Generally							
Substructure	115	14	78	104	138	470	1,003
Superstructure	489	48	382	461	576	1,491	1,003
Internal Finishes	110	12	85	105	130	365	1,003
Fittings	62	0	33	53	79	499	994
Services	327	42	235	311	396	1,419	1,004
712. Primary Schools							
Generally							
Substructure	115	26	83	105	138	470	462
Superstructure	474	190	378	447	552	1,101	462
Internal Finishes	110	14	89	108	131	268	462
Fittings	58	4	36	51	75	232	457
Services	319	42	240	305	380	990	462
712. Primary Schools: Up to 500 m² GFA							
Substructure	143	64	103	144	170	261	67
Superstructure	554	194	423	485	665	1,101	67
Internal Finishes	122	30	98	119	141	268	67
Fittings	62	10	40	57	86	150	64
Services	358	137	264	342	426	990	67
712. Primary Schools: 500 m² to 2,000 m² GFA							
Substructure	109	26	81	101	124	470	338
Superstructure	454	190	369	439	519	866	338
Internal Finishes	106	14	85	103	126	215	338
Fittings	56	4	35	49	72	232	337
Services	308	42	234	290	367	663	338

(Continued)

Table 7.3 (*Continued*)

Building Function	£/m² Gross Internal Floor Area						Sample
	Mean	Lowest	Lower Quartile	Median	Upper Quartile	Highest	
712. Primary Schools: Over 2,000 m² GFA							
Substructure	121	36	86	112	145	257	57
Superstructure	497	205	403	481	569	1,083	57
Internal Finishes	119	67	102	113	134	204	57
Fittings	60	8	40	57	78	165	56
Services	340	170	257	323	406	705	57
713. Secondary Schools (High Schools)							
Substructure	98	14	59	86	119	438	168
Superstructure	457	222	347	430	525	1,332	167
Internal Finishes	100	34	76	97	111	365	167
Fittings	57	0	28	52	79	224	166
Services	286	84	205	272	337	1,419	168

Note: *Unpriced Exclusions* column deleted for clarity.
Source: BCIS Online.

To show the comprehensiveness of the BCIS database for this particular type of building (and the same applies to the other categories) the full list of school types is as follows:

- Schools:
  - generally
  - public
  - private.

- Nursery schools/crèches:
  - generally
  - up to 500 m² GFA
  - 500–2,000 m² GFA
  - over 2,000 m² GFA.

- Primary schools:
  - generally
  - up to 500 m² GFA
  - 500–2,000 m² GFA
  - over 2,000 m² GFA.

- Middle schools

- Primary/middle – specialised teaching block

- Primary schools – mixed facilities
- Secondary schools (high schools)
- Secondary schools – specialised teaching blocks.

The whole list of rates per group element for these has been excluded, as the record would be too extensive for ease of reading.

PUBLISHED SOURCE

A common alternative to the above approach where a suitable cost analysis is not available is to use a published source, such as *Spons*, *Laxtons* in the UK or in Australia, *Rawlinsons Australian Construction Handbook* (current handbook). This publication contains a section where elemental breakdowns are given in the part titled, 'Estimating – Elemental Costs of Buildings'.

A sample page from this section is given on the following page in Figure 7.4.

You should select the elemental breakdown of the project type with the brief description that most closely matches the new project. These costs are expressed as $ per square metre and are broken down into the various group and individual elements. Of great importance to the cost planner is the column showing the percentage of the total building cost for every building. This percentage provides the key to breaking down the budget into the group and Individual elements. The use of this column will be demonstrated later.

Note that External Services in these cost analyses represent only a small amount of the total cost because these are the services taken from the external wall to a distance of 3.0 m around the perimeter of the external wall. The remainder of the External Services from this 3.0 m distance from the external wall to the boundary would have to be assessed for each site individually and as a separate item in the cost plan.

DERIVING COST TARGETS

The total target cost for the project is obtained by updating the cost per square metre of GFA, and multiplying the result by the total GFA of the new project. Once the updated target cost has been calculated by this means, then the group element cost targets can be derived by applying the appropriate percentage from the cost analysis, range of cost analyses or from the elemental breakdown in the chosen publication.

Example

So, for instance if we have a 6-storey office block with a GFA of 10,000 m², using Figure 7.4 as a guide, we can break down our overall cost target into group and elemental cost targets. Note that for simplicity, the rate of

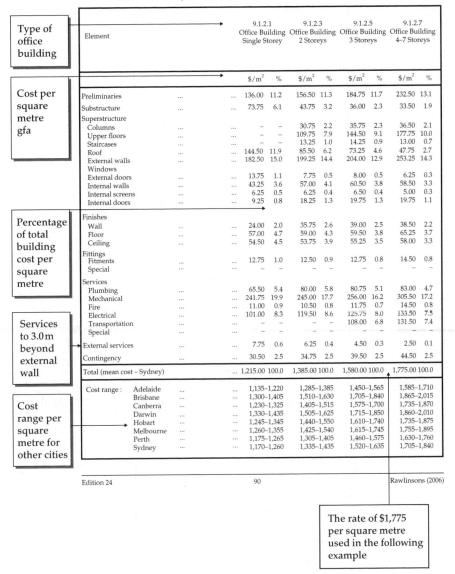

Estimating: Elemental costs of buildings

9.0 **Offices**

9.1 Lettable low rise (fully serviced) – Finished Floor Lettable

Element			9.1.2.1 Office Building Single Storey		9.1.2.3 Office Building 2 Storeys		9.1.2.5 Office Building 3 Storeys		9.1.2.7 Office Building 4–7 Storeys	
			$/m^2	%	$/m^2	%	$/m^2	%	$/m^2	%
Preliminaries	...	...	136.00	11.2	156.50	11.3	184.75	11.7	232.50	13.1
Substructure	...	...	73.75	6.1	43.75	3.2	36.00	2.3	33.50	1.9
Superstructure										
Columns	...	...	–	–	30.75	2.2	35.75	2.3	36.50	2.1
Upper floors	...	...	–	–	109.75	7.9	144.50	9.1	177.75	10.0
Staircases	...	...	–	–	13.25	1.0	14.25	0.9	13.00	0.7
Roof	...	...	144.50	11.9	85.50	6.2	73.25	4.6	47.75	2.7
External walls	...	...	182.50	15.0	199.25	14.4	204.00	12.9	253.25	14.3
Windows										
External doors	...	...	13.75	1.1	7.75	0.5	8.00	0.5	6.25	0.3
Internal walls	...	...	43.25	3.6	57.00	4.1	60.50	3.8	58.50	3.3
Internal screens	...	...	6.25	0.5	6.25	0.4	6.50	0.4	5.00	0.3
Internal doors	...	...	9.25	0.8	18.25	1.3	19.75	1.3	19.75	1.1
Finishes										
Wall	...	...	24.00	2.0	35.75	2.6	39.00	2.5	38.50	2.2
Floor	...	...	57.00	4.7	59.00	4.3	59.50	3.8	65.25	3.7
Ceiling	...	...	54.50	4.5	53.75	3.9	55.25	3.5	58.00	3.3
Fittings										
Fitments	...	...	12.75	1.0	12.50	0.9	12.75	0.8	14.50	0.8
Special	...		–	–	–	–	–	–	–	–
Services										
Plumbing	...	...	65.50	5.4	80.00	5.8	80.75	5.1	83.00	4.7
Mechanical	...	...	241.75	19.9	245.00	17.7	256.00	16.2	305.50	17.2
Fire	...	...	11.00	0.9	10.50	0.8	11.75	0.7	14.50	0.8
Electrical	...	...	101.00	8.3	119.50	8.6	125.75	8.0	133.50	7.5
Transportation	...	...	–	–	–	–	108.00	6.8	131.50	7.4
Special	...						–	–	–	–
External services	...	...	7.75	0.6	6.25	0.4	4.50	0.3	2.50	0.1
Contingency	...		30.50	2.5	34.75	2.5	39.50	2.5	44.50	2.5
Total (mean cost – Sydney)	...		1,215.00	100.0	1,385.00	100.0	1,580.00	100.0	1,775.00	100.0
Cost range : Adelaide	...	...	1,135–1,220		1,285–1,385		1,450–1,565		1,585–1,710	
Brisbane			1,300–1,405		1,510–1,630		1,705–1,840		1,865–2,015	
Canberra	...		1,230–1,325		1,405–1,515		1,575–1,700		1,735–1,870	
Darwin	...		1,330–1,435		1,505–1,625		1,715–1,850		1,860–2,010	
Hobart	...		1,245–1,345		1,440–1,550		1,610–1,740		1,735–1,875	
Melbourne	...		1,260–1,355		1,425–1,540		1,615–1,745		1,755–1,895	
Perth	...		1,175–1,265		1,305–1,405		1,460–1,575		1,630–1,760	
Sydney	...		1,170–1,260		1,335–1,435		1,520–1,635		1,705–1,840	

Labels on the left margin pointing to the table:
- Type of office building
- Cost per square metre gfa
- Percentage of total building cost per square metre
- Services to 3.0 m beyond external wall
- Cost range per square metre for other cities

Edition 24   90   Rawlinsons (2006)

> The rate of $1,775 per square metre used in the following example

Figure 7.4 Elemental Breakdowns for Estimating

$1,775/m^2$ for the office building, 4–7 storeys (9.1.2.7) has been converted directly into the same £ sterling amount.

Target tender cost $= 10,000\,m^2 \times \$1,775/m^2$ ($= £1,775/m^2$) (as Figure 7.4, building 9.1.2.7) $= £17,150,000$. The cost breakdown is shown in Table 7.4.

Table 7.4 Office Cost Plan: Cost Breakdown

Group Element	Individual Element	Group (%)	Elemental (%)	Cost Allocation (Target Cost £17.15 m)
Preliminaries		13.1	13.1	£2,246,650
Substructure		1.7	1.7	£325,850
Superstructure		34.8		
	Columns		2.1	£360,150
	Upper floors		10.0	£1,715,000
	Staircases		0.7	£120,050
	Roof		2.7	£463,050
	External walls		14.3	£2,452,450
	Windows		In external walls	
	External doors		0.3	£51,540
	Internal walls		3.3	£565,950
	Internal screens		0.3	£51,540
	Internal doors		1.1	£188,650
Finishes		9.2		
	Wall		2.2	£377,300
	Floor		3.7	£634,550
	Ceiling		3.3	£565,950
Fittings		0.8		
	Fitments		0.8	£137,2000
	Special		–	
Services		37.6		
	Plumbing		4.7	£806,050
	Mechanical		17.2	£2,949,800
	Fire		0.8	£137,200
	Electrical		7.5	£1,286,250
	Transportation		7.4	£1,269,100
	Special		–	
External Services		0.1		£17,150
Contingency		2.5		£428,750
Totals		100	100	£17,150,180

Through the simple device of taking an existing cost model of a representative and similar building, whether from the BCIS cost database (or from published sources as above), we can realistically set group elemental targets (at least) and be reasonably confident that they represent practical cost targets for our cost planning.

Other costs
In addition to the cost of the building(s) the costs of all work beyond the external walls of the building(s) must also be included so that the total cost reported to the client is inclusive of all the items needed to complete the building ready for handover to the client. The extent and scope of these costs will largely depend on the project site; its area and characteristics. Jaggar et al. (2002: 74) have identified the factors that can influence the additional costs that need to be taken account of in the cost plan. These factors are:

- Topography of the site (flat, sloping, undulating and natural features).

- Proximity to existing buildings and structures (built-up city site).

- Availability of existing site services (remoteness).

- Greenfields or brownfields sites can bring problems or advantages with each type.

- Contaminated land and the need to remove the contamination totally from the site.

- Specific planning restrictions such as height, density, setbacks, site coverage, heritage requirements.

- Proximity to the range of utilities (gas, water, electricity, sewerage, drainage).

- Site restrictions such as parking, noise levels, overlooking, shadows, glare.

The need for measures to provide or mitigate these factors needs careful consideration by the design team and cost planner. The cost of these items must be carefully researched by the cost planner even though detailed information is scarce or totally absent at this stage. Previous experience or discussions with colleagues who may have information on the particular item are likely to be required, but eventually when an amount is decided it may be just the best guess until more information becomes available.

SUGGESTED APPROACH TO THE ASSESSMENT OF PRELIMINARIES

The adjustment of preliminaries deserves individual treatment. It is an element that has received scant attention in the cost planning literature, but it is essential that the cost planner fully understand the nature of the element and the factors that affect its level of pricing. The accurate pricing of preliminaries based on earlier projects can be difficult due to the following problems:

(a) Preliminaries are more directly influenced by the choice of construction method than any other element. The site organisation and method of construction are generally not known (nor even considered) at the time the cost plan is prepared, since it is a decision made by the contractor who has not been appointed yet. This makes the assessment of preliminaries costs difficult to accurately forecast. The use of cost analyses of previous projects provides the best guide to pricing this element.

(b) The examination of preliminaries cost in a number of previous projects was undertaken by Gray (1980) and Flanagan (1980) in unpublished research theses. The results from this work illustrate the variability of this element, and are shown in Table 7.5.

(c) The preliminaries section in the tender is often used by management in the contracting firm to make last-minute changes to the tender bid. Consequently, the adjusted preliminaries section often then does not reflect nor represent the true cost of this section of the project. These arbitrary changes make the analysis of preliminaries difficult in practice and lead to problems of forecasting their true cost.

The percentage of the cost attributable to preliminaries items can vary considerably between different projects and for no easily identifiable reason. In Table 7.5 above the mean percentage on all the projects analysed varies between 16.5% and 33.0%. These are considerable differences, which can create problems in accurately pricing the preliminaries element, and can contribute to significant errors in the forecast total cost.

Preliminaries assessed at the design stage of a project are most commonly calculated as a percentage of the total of the remainder of the work. That is, the total cost of elements 1. Substructure to 6. External Works and excluding preliminaries.

Table 7.5 Variability in Preliminaries Costs

Project Value Range (£ ,000)	Number in Range	Preliminaries Range (%)	Mean	Coefficient of Variation
0–100	47	12.6–64.4	33.0	38.5
100–1,000	114	12.6–54.6	25.4	32.3
1,000–5,000	53	12.6–58.8	18.3	40.4
5,000–10,000	2	15.4–19.6	16.5	15.2

Source: Property Services Agency (1981: 27) Cost Planning and Computers, Department of the Environment, London, UK.

FACTORS TO CONSIDER

The percentage included for preliminaries is often considered in a cursory fashion, usually based on previous projects and an interpretation of information available on the project being cost planned.

However, the cost planner should consider the following range of items and factors that can greatly influence the percentage allowed for the preliminaries:

(a) *Location*: Assessment of ease or difficulty of entry to and egress from the site, distance from major highways, requirements for temporary roads.

(b) *Space available on site*: Limited on-site storage and space for facilities (inner city sites) is likely to require special measures leading to increased costs.

(c) *Security*: Temporary fencing, hoardings and gantries required for public safety and to protect the works from pilfering and vandalism are significant costs affected by the location and length of the boundary of the site.

(d) *Contract period*: Many items in preliminaries are related directly to the length of the construction period – supervision, cleaning, accomm-odation, hire, telephones and the like. Additionally, when a client requires an early completion time the preliminaries section may be increased to account for overtime payments, weekend working and penalty rates generally.

(e) *Plant*: Many items of plant, particularly those provided by the subcontractor, are directly related to the works in an element and are priced in that section. However, where the contractor provides general items of plant for use by the whole site and across several subcontractors then these are likely to be priced in the preliminaries element. Examples of these include scaffolding, hoists, site cranes and a range of small tools.

(f) *Insurance*: The requirement for insurance for public liability, adjoining owners and value of the works is included in preliminaries and should be judged on the basis of risks influencing the specific site.

A comprehensive listing of all the items that should be included in the pre-liminaries element is given in the *Standard method of Measurement*, 7th edition, (Royal Institution of Chartered Surveyors (RICS), 1989) (amended 2000) and the Australian *Standard Method of Measurement*, 5th edition (Joint Standard Method of Measurement Committee, 1990).

Sadly, when past projects are being analysed it is rare that a breakdown of the value of the items making up the preliminaries is given. Most commonly the total cost of preliminaries is given without any further information. This makes the assessment of preliminaries in future projects more difficult. Guidance on the pricing of some of the major preliminaries items is contained in most of the published cost guides.

Cost Indices

A review of the BCIS Tender Price Index and the Market Conditions Index should provide a sound guide for the cost planner to make an assessment of the present date cost of the project and evaluate the possible price changes to the tender date and for fluctuations during the construction period. See Chapter 5 for a discussion of the BCIS indices and the price and cost movements they measure.

If the BCIS is not used and the automatic indexing facility cannot be used, then the costs used for the Outline Proposals cost plan must be updated in three stages. Firstly, to the date of the cost plan. Next, the present date costs must be projected forward to the expected tender date. Finally, the tender costs are projected forward for the length of the construction period to take account of the expected fluctuation in costs that the contractor will either claim for, or add to the tender sum to make it a fixed price contract. These updating stages are shown diagrammatically in Figure 7.5.

Completing the Cost Plan

During the preparation of the outline cost plan, the various allowances should obviously be made with an eye on the overall cost limit. If the cost limit is exceeded by all the alternative proposals then the design team must review the situation, make any changes and recommendations as necessary and report the outcome to the client. However, if any of the alternatives exceed the cost limit then this must be clearly noted in a report to the client. It should not be assumed that when an alternative exceeds the cost limit that it is automatically excluded from consideration because the extra costs may provide a much bigger dividend in extra value gained for the project as we discussed in Chapter 1.

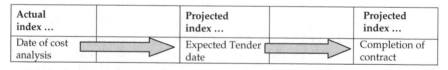

Actual index ...		Projected index ...		Projected index ...
Date of cost analysis		Expected Tender date		Completion of contract

Figure 7.5 Updating Costs

When the Outline Proposals cost plans have been completed, they are sent with the favoured outline design for the client's approval.

PREPARING THE OUTLINE COST PLAN

To maintain strict control over costs, the ideal solution is obviously for the designing and costing to be done at the same time. Moreover, there should be considerable interaction between the two, with every cost decision being allowed to affect the corresponding design decisions, and vice-versa.

This is obviously reaching the height of perfection, but it is an aim that the design team should aspire to. It should be noted that for clarity and simplicity we have shown the design decisions as being completed before the outline cost plan is started. We have done this only to eliminate complicated descriptions of the timing and the complex decision-making of a dynamic Outline Proposal stage.

A format is suggested for presenting the outline cost plan. This format is largely based on the AIQS (2000), as it is one of the few publications that provide practical guidance on cost plan presentations.

STAGE C: OUTLINE PROPOSALS COST PLAN

Proposed project details: Outline information
Primary school, single storey: 2,150 m² GFA.
New build
UK mean location
Expected quality – medium to high
Cost per square metre GFA = £1,400.00 building only.
Cost limit: £4.0 million
Present date: December 2005
Tender date: July 2006
Completion date: July 2007
Allowances: Site works (£230,000); External Services (£55,000); Removal of asbestos waste (£105,000).

Background

First of all we interrogate the BCIS Online Cost database for new build schools, then primary schools, next schools over 2,000 m². To show how the updating of project costs without the automatic updating and indexing system of the BCIS we are going to 'manually' update the project costs as an example of the process noted above. The BCIS index for our purposes is indexed at the fourth quarter of 2005 and based on our UK mean location. The extract from the BCIS is given in Table 7.6 with the specific category of costs shown.

Table 7.6 BCIS Group Element Prices

Group Element Cost per Square Metre Study							
Rate per square metre gross Internal floor area for the group element excluding external works and contingencies and with preliminaries apportioned by cost. Last Updated 1 April 2006. At 4Q2005 prices (based on a Tender Price Index of 223) and UK mean location.							
Building Function	$£/m^2$ Gross Internal Floor Area						Sample
	Mean	Lowest	Lower Quartile	Median	Upper Quartile	Highest	
New Build							
Over 2,000 m² GFA							
Substructure	121	36	86	112	145	257	57
Superstructure	497	205	403	481	569	1,083	57
Internal Finishes	119	67	102	113	134	204	57
Fittings	60	8	40	57	78	165	56
Services	340	170	257	323	406	705	57

SELECTION OF A SUITABLE RATE PER SQUARE METRE FOR GROUP ELEMENTS

We expect this primary school to be above average in quality as noted above and so we shall select the upper quartile costs from the BCIS data in Table 7.6. The locality index for the south-east of England according to the BCIS indices is +8% above the UK mean. The cost allocation to each group element is based on the costs per square metre given in Table 7.7 with the total cost for each group element derived by multiplying this costs per square metre by the GFA of 2,150 m². The results are shown in Table 7.7.

The allowance for preliminaries in this project is not a separate item as the costs per square metre used in this cost plan are inclusive of the preliminaries element. If these costs per square metre were not inclusive of a suitable global preliminaries allowance then a separate allowance for preliminaries would be made for Item 6 on the cost plan.

The allowances for various works beyond the building are also included. The planning, design (risk) and locality allowances all reflect the type of information available at this stage and the relatively long time needed for the project to come to completion.

The BCIS index (BCIS All-In Tender Price Index) is used to provide the means for adjusting the costs to tender (escalation to tender = 8 months) and completion date (escalation during contract period = 12 months) and these allowances are shown as a manual percentage calculation in Table 7.7. This

Table 7.7 Outline Cost Plan Calculations

Group Elements	Building Cost (%)	£/m² GFA	Total Cost (£)
1. Substructure	10.9	145	311,750
2. Superstructure	42.7	569	1,223,350
3. Finishes	10.1	134	288,100
4. Fittings and furnishings	5.9	78	167,700
5. Services	30.4	406	872,900
6. Preliminaries	Included	–	–
Building cost totals	100.0	£1,332	2,863,800
Project allowances			
• Alterations	Nil		–
• Site works	Estimate		230,000
• External Services	Estimate		55,000
• Removal of asbestos waste	Estimate		105,000
• Other allowances as required			
Totals			3,253,800
Contingencies/allowances			
• Planning	+2%		65,080
• Design	+5%		162,690
Subtotal			3,481,570
• Locality	+8%		278,530
Net project cost			3,760,100
Special provisions			
• None identified			—
Allowance for contract contingencies	+3%		112,800
Project cost – Present date (4Q2005 = Index 223)			3,872,900
Escalation to tender date (8 months)	+3.6%		139,420
Project cost – tender date (3Q2006 = Forecast 231) 223/231			4,012,320
Escalation during contract (12 months) (3Q2007 = Forecast 241) 231/241	+4.3%		175,530
Total project commitment			£4,184,850
Comparison with cost limit			
• Cost limit		£4,000,000	
• Total project commitment		£4,184,850	
Overexpenditure	+4.6%	£184,850	—
~~Underexpenditure~~			–

adjustment can also be carried out automatically by the BCIS, but we wished to illustrate the details of these adjustments to show how they are carried out.

The additional allowances for the various contingencies of planning, design and contract are also shown. An assessment of the planning allowance of +2% represents a remaining small concern that the planning approvals and possibly the programme could be delayed. The design risk of +5% is for an allowance for increased costs due to more research and work required in the design and detailing of the project. Contract contingencies of +3% represent an assessment of this amount at this early stage and will reduce to the actual value of the contract contingencies to be included in the project tender documentation. The locality index value of +8% above the mean UK location has been gained from the BCIS.

The Outcome

This overexpenditure of £184,850 at this stage is a warning to all the design team that they must all be cost conscious. However, such a small overexpenditure (less than 5%) is not a reason to review or cancel the whole project.

The results of the cost planning should be clearly shown for each alternative proposal and the comparison with the cost limit or budget is essential information for the design team and the client.

For all cost plans it is also advisable to include a list of exclusions for client and other parties to make certain that everyone knows what the costs presented include. A typical list of exclusions is given here for guidance, not prescription.

The following items have been EXCLUDED from this cost plan:

- Value added tax (VAT).
- Allowances for escalation in costs from beyond the projected completion date of the works.
- Any land costs or land acquisition expenses, legal fees, planning fees, survey and building permit or inspection fees and the like.
- Professional, project management and any other consultant fees.
- Specialist advisors and the work they may recommend.
- Loose furniture and any equipment.

Examples of Outline Proposals cost plans are also presented in Flanagan and Tate (1997) and Smith and Love (2002).

If the cost limit is not breached and an alternative has been chosen or recommended by the design team and approved by the client the project can proceed with confidence to the next stage, Detailed Proposals. If the budget or cost limit has been breached by the proposal(s) then the design team is in for more work to bring it within budget or to convince the client to increase the budget if it is not possible to bring the project back within the original budget.

Chapter Review

The Outline Proposals stage begins to expand the cost information base of the project. The single figure budget or cost limit provided at the Strategic Briefing stage (or earlier) is developed into a skeleton of cost allocations in the form of group elemental costs. These group costs should be a realistic assessment of costs for that particular project type that forms a sound basis for the design team to work within to produce a balanced design to satisfy the client's requirements. Ideally, the end of this stage should confirm the budget. Alternatively, a more realistic cost assessment will have been established to revise the cost limit for the remaining stages.

At the end of the Outline Proposals stage, the major cost and quality parameters of the project will have been largely established and the opportunity for change in the subsequent stages will be much more limited. Any major changes after this crucial stage could result in a great deal of wasted effort and redesign of part, or the whole, of the project. As noted in earlier chapters, possibly up to 75% of the total project costs will have been committed by the end of the Strategic Briefing and Outline Proposals stage. This reinforces the view that the design team should work cooperatively and effectively as a team to review rigorously all the major design decisions and alternative proposals taken during the Strategic Briefing and Outline Proposals stages. Failure to do so could have a negative impact on the final outcome. At worst, it could lead later in the process to the postponement or cancellation of the project.

It is important to remember that in neither the Strategic Briefing nor the Outline Proposals stages has any commitment been made to a particular specification or design choice. The only decision that can be considered to be fixed is that the total budget is not exceeded, or there is a good reason to modify the cost limit. The design team still has ample freedom within a realistic budget to consider a wide range of choices. At the Outline Proposals stage the group elemental costs do not provide an inflexible cost limit for arriving at a satisfactory design solution. They are sound, balanced cost allocations based on previous similar projects and provide a reasonable basis for making design decisions for the project in hand. It is also important that the client is fully aware of the status of the cost information provided at the Brief and Outline Proposals stages and not to expect a finality of design or accuracy of costing that cannot be guaranteed at these stages.

SUMMARY OF COST PLANNING DURING OUTLINE PROPOSALS

- Cost targets are prepared for each group of elements by allowing for the major differences between the building(s) analysed from the BCIS (or from published sources) and the new project, then updating them to the date of preparation of the outline cost plan.

- An allowance is made for planning contingencies, design contingencies, contract contingencies, project contingencies, escalation and contract cost adjustments. These are made by assessing the cost of unforeseen design difficulties and possible price rises between the preparation of the outline cost plan, the receipt of the tender and during the contract period.

If there is a major discrepancy between the outline cost plan and the cost limit, it is best for the client and the design team to make a *joint* decision as to whether it is better to adjust the allowances, or to accept that the outline cost plan total is more realistic than the first estimate prepared during the Strategic Briefing stage). If this results in the cost limit being raised rather than lowered, at least the client has played an active part in the decision and has been afforded the opportunity of abandoning the project before incurring major expenditure.

When the client accepts the outline cost plan the design team is committed to producing the Detailed Proposals (Sketch or Scheme Design) within the cost shown on the outline cost plan.

Review Question

How is an Outline Cost Plan prepared?

What are the cost allocations and allowances that are included in an Outline Proposals cost plan?

Give brief details of how they are prepared.

References

AIQS (2000) *Cost Control Manual*, National Public Works Council Inc., Canberra.

BCIS (1969, reprinted 2003) *Standard Form of Cost Analysis*. The Royal Institution of Chartered Surveyors, London, UK.

BCIS Online (1998) http://www.bcis.co.uk

Brown, G. and Matysiak, G. (2000) *Real Estate Investment: A Capital Market Approach*, Harlow: Prentice-Hall Financial Times, Chapter 7: Valuations and prices.

Cadman, D. and Topping, R. (1995) *Property Development*, 4th edition, E&FN Spon, London, UK.

Flanagan, R. (1980) *Tender Price and Time Prediction for Construction Work*, PhD thesis, unpublished, University of Astron, Birmingham, UK.

Flanagan, R. and Tate, B. (1997) *Cost Control in Building Design*, Blackwell Science, Oxford.

Gray, C. (1980) *Analysis of the Preliminary Element of Building Production Costs*, MPhil thesis, unpublished, University of Reading, UK.

Isaac, D. (1996) *Property Development: Appraisal and Finance*, Palgrave, Basingstoke, UK.

Jaggar, D., Ross, A., Smith, J. and Love, P. (2002) *Building Design Cost Management*, Blackwell Publishing, Oxford.

Joint Standard Method of Measurement Committee (1990) *Australian Standard Method of Measurement of Building Works*, Australian Institute of Quantity Surveyors and Master Builders'-Construction and Housing Association Australia, Inc., Deakin, Australian Capital Territory, Australia.

Junior Organisation Quantity Surveyors Standing Committee (1976) *An Introduction to Cost Planning*, The Royal Institution of Chartered Surveyors, London, UK.

Laxtons Building Price Book (2006) Johnson, V. B. (editor), Elsevier Press, Oxford.

Ministry of Public Buildings and Works (1968) *Cost Control during Building Design*, HMSO, London, UK.

Property Services Agency (1981) *Cost Planning and Computers*, Department of Environment, London.

Rawlinsons (annual publication) (2006) *Rawlinsons Australian Construction Handbook*, Rawlhouse Publishing, Perth, Western Australia, Australia.

Robinson, J. (1989) *Property Valuation and Investment Analysis: A Cash Flow Approach*, Law Book Company, Sydney.

Royal Institution of British Architects (2000) *The Architect's Plan of Work*, RIBA Publications, London, UK.

RICS (1989) (amended 2000) *Standard method of Measurement*, 7th Edition, The Royal Institution of Chartered Surveyors, London, UK.

Smith, J. and Love, P. (2000) *Building Cost Planning in Action*, University of New South Wales Press, Sydney, Australia.

Spon's Architect's and Builders' Price Book (2006) Davis Langdon (editor), E&FN Spon, London, UK.

Further Reading

Bathurst, P. E. and Butler, D. A. (1983) *Building Cost Control Techniques and Economics*, Heinemann, London, UK.
    Chapter 17: Cost planning: application, pp. 103–124.

Building Economist, any volume, refer to the Cost Index, The Australian Institute of Quantity Surveyors, Canberra.

Cartlidge, D. P. and Mehrtens, I. N. (1982) *Practical Cost Planning*, Hutchinson, London, UK.
    Chapter 7: Cost planning during the outline proposals and scheme design stages, pp. 77–89.

Flanagan, R. and Tate, B. (1997) *Cost Control in Building Design*, Blackwell Science, Oxford.
    Chapter 5: Cost control during inception, feasibility and outline proposals, pp. 49–56.
    Chapter 10: Cost planning during inception, feasibility, outline proposals and scheme design, pp. 187–201.
    Chapter 11: Cost example from feasibility to scheme design, pp. 243–254.

SUGGESTED ANSWERS

Question 7.1: Outline Cost Plan Allowances

Outline cost plan of a simple office building
Assume date for Outline Proposals stage is the fourth quarter 2005.
Our discussion below will relate to the general state of the market at this time.

The floor area is approximately 2,200 m².

Briefing Stage budget = £1,870,000

The first calculation we make is to calculate the cost per square metre GFA.

$$\text{Total cost per square metre} = \frac{£1,870,000}{2,200\,\text{m}^2} = £850/\text{m}^2$$

- *Design contingency*
  Since our office block is a fairly simple project, we decide that an allowance of 3.5% is a large enough margin for any unforeseen eventualities in the development of the design. Thus, 3.5% is judged to be an adequate amount for design contingencies.

- *Contract contingency*
  There are a number of possible unforeseen problems at this early stage, so we shall allow a percentage in the higher range of 4.0%.

- *Escalation to tender date*
  Let us assume construction activity is steady, but not buoyant. We therefore, anticipate a low price increase for the period we are considering. An allowance of +1.0% is made to cover the normal fluctuations in such a competitive market. We would also consult with the BCIS Online Cost indices and judge the forecasts made in these indices for this type of building in this locality.

  We will assume a contract period of 48 weeks and a relatively stable period of costs where the successful contractor is likely to include a low estimate of future price increases in the fixed price contract. An allowance of +1.0% is made to cover contract adjustments during the contract period of 48 weeks. Again, we would also consult with the BCIS Online Cost indices and judge the forecasts made in these indices for the period of time required to complete the contract. Also note that the forecasts in these indices only usually go forward for just over 2 years.

Summary of Allowances

This gives a total sum of the allowances as:
- Design contingency   3.5%
- Contract contingency   4.0%
- Escalation   1.0%
- Contract cost adjustment   1.0%

Total allowance   9.5%

Sums adding to totals of nearly 10% in a relatively depressed market cannot be ignored and alert the cost planner to the fact that to ignore considerations of these allowances is dangerous. They can make a substantial difference to the accuracy of final cost plan estimates.

$$\text{Cost allowance in our project} = £850 \times \frac{9.5}{100} = £80.75/\text{m}^2$$

Thus, the total cost per square metre is £930.75/m².

Total cost to end of the project = 2,200 m² × £930.75/m² = £2,047,650.

Review Question: How is an Outline Cost Plan Prepared?

The objective of the Outline Proposals stage is to allocate the available money to the various parts of the proposed building, without making any major design decisions, which may be impossible to maintain at a later stage. Major differences are certainly isolated and allowed for, but if this was all that was done, the only action that could follow would be to confirm or amend the cost limit.

However, there is an important distinction between an outline cost plan and a mere tabulation of allowances for the major project differences that become apparent at Outline Proposals. A cost target is prepared for each group of elements. This has the advantage that it forces the design team to think clearly at this early stage about the broad allocation and balance of expenditure to the major parts of the building.

The first step in preparing the cost target for a group of elements is to identify the costs of previous projects that bear the strongest similarity to the project in hand. With access to the BCIS data the cost planner has the ability to tap into a range of projects and for the example primary school project we used earlier in the chapter there were 57 similar projects we had access to. So, the costs per square metre of each group element provided an excellent cost base for our project cost plan.

The main basis for selecting an appropriate cost model to provide the basis of the cost plan allowances is the list given earlier as follows:

- function(s) or use(s),
- new build or renovations,
- area or space,
- number of storeys,
- location,
- general statement of quality,
- type of construction.

Another important adjustment is the one for differences in time between the source cost data and the tender date for the new project and then to completion of the project. The other contingency allowances as follows also need to be given some serious consideration:

- Planning contingency
- Design (risk) contingency
- Contract contingency
- Project contingency.

The allowances are more detailed than those made during the Strategic Briefing stage and the cost detail begins to reflect the type of information being produced by the design team. The development of costing the project should be seen as a continuing and evolving process where greater accuracy and confidence in costing occurs as the project progresses.

Chapter 8

Cost Planning the Detailed Proposals Stage

> The traditional view of cost planning is that it moves through states of increasing accuracy beginning with fairly low accuracy at the inception stage and then as the sketch designs are produced elemental cost planning takes place and eventually cost plans from approximate to firm quantities.
>
> Raftery (1985: 62)

Chapter preview

Now we get down to developing the detail of our project information in the form of more detailed plans (this stage was previously termed, sketch plans or scheme design) now begins to flow and probably for the first time we can now start to estimate costs and cost plan on the basis of the detail now contained in our project rather than taken from similar projects. The exchange of views and opinions between the cost planner(s) and the design team can begin in earnest presenting a challenge of communication, professional expertise, expediency, resilience, tact and persuasiveness on each member's part. It is a demanding stage requiring a high level of personal skills, but the aim is always to gain the best building for the client within the resources that are available. Relaxing the grip of cost planning and accepting without question decisions made by the design team or the client has no role in this crucial stage.

Having completed the Strategic Briefing stage B and Outline Proposals stage C, the next stage in the process we shall focus on is the Detailed Proposals (stage D). As in the previous stage of Outline Proposals, it is best to start preparing the detailed cost plan as early as possible during this stage. The cost targets established during these early stages can be checked as the details of the project emerge in the form of sketch plan drawings and other information as it is completed.

This chapter continues with the theoretical and applied aspects of earlier chapters. The Strategic Brief and Outline Proposals stages that were the focus of our attention in Chapters 6 and 7 are now complete. Using the framework diagram we introduced earlier, the Detailed Proposals stage is shown shaded in Table 8.1.

In this chapter we will again review the main theoretical ideas (or three principles) underlying the approach to cost planning activities. These are

Table 8.1 RIBA Outline Plan of Work 1998

Pre-design (Briefing)			Design			
Inception or Feasibility			Pre-construction Period			
Pre-stage A	Work Stage A	Work Stage B	Work Stage C	Work Stage D	Work Stage E	Work Stage F
	1	2	3	4	5	6
Establishing the Need	Options Appraisal	Strategic Briefing	Outline Proposals	Detailed Proposals	Final Proposals	Production Information
Chapter 6			Chapter 7	Chapter 8	Chapter 9	
Briefing			Outline Proposals	Detailed Proposals	Documentation	

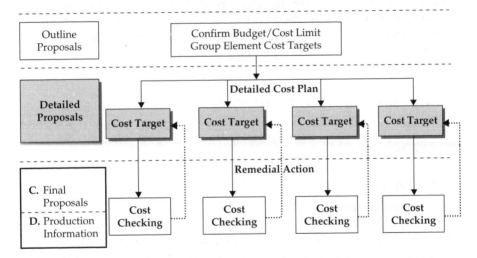

Figure 8.1 Cost Planning Concept Map: Detailed Proposals stage
Source: Adapted from Smith and Love (2000: 11).

summarised in the concept map of Figure 8.1 where the key role of this stage in converting the budget (or cost limit) into a working cost plan with realistic cost targets established is shown. We would like to emphasise again that the method of approach described in this chapter also attempts to simulate good practice.

However, the cost planner should not become complacent by applying the same approach or a single prescriptive formula in every situation. Good judgement and skill still have to be applied in each new project environment and this material provides a guide, not a fixed track, to a single solution. A reflective rather than a reactionary approach accepting of new ideas and innovatory solutions should be adopted in all projects and problem solving situations such as these design stages.

Thus, the information available when the cost plan is started is:

- the project brief (probably not fully complete) from the Outline Proposals stage;

- the outline design (including drawings);

- the outline development control approval from the planning authority;

- the outline cost plan to group element level; and

- all the information created during and at the end of the Outline Proposals stage.

The purpose of this stage is to ensure that the overall design is the most effective available in terms of the approved requirements as expressed in the project brief and other documents noted above. The cost planning activities should confirm (and only in rare circumstances set the final budget) and more importantly from the cost planning point of view, establish elemental cost targets.

BUDGET DISTRIBUTING TECHNIQUES

As noted in the previous two chapters there are two basic types of cost planning technique:

- budget setting and

- budget distributing.

At the Detailed Proposals stage we are definitely in budget distributing territory and we can identify this by reference to the diagram we first used in Chapter 6. This is shown in Figure 8.2 where budget distributing techniques and the stages they are used are clearly identified.

Elemental Method (Individual Elements)

This method relies upon the availability of suitable cost analyses and other cost data together with an elemental breakdown of the estimated project. It also relies upon the use of approximate quantities and a range of ratio and other adjustments for quantity and quality adjustments from the cost databases, analysis (or analyses) to the cost plan.

The standard elements for all buildings in the UK have been defined by the RICS (1969) in the *Standard Form of Cost Analysis*. A copy of this is contained in the Appendix. The standard list of elements and units of measurement from this standard form are summarised in Table 8.2.

Pre-design			Design			
Inception or Feasibility			Pre-construction Period			
Pre-stage A	Work Stage A	Work Stage B	Work Stage C	Work Stage D	Work Stage E	Work Stage F
	1	2	3	4	5	6
Establishing the Need	Options Appraisal	Strategic Briefing	Outline Proposals	Detailed Proposals	Final Proposals	Production Information
EARLY STAGE TECHNIQUES (Described later in Chapter) • Strategic Value Management • Strategic Needs Analysis • Quality Function Deployment • Situation Structuring		Cost Limit	Cost Targets		Cost Checking	
	• Expert Choice					
	BUDGET SETTING TECHNIQUES		BUDGET DISTRIBUTING TECHNIQUES: BUDGET DISTRIBUTION, CHECKING AND RECONCILIATION			
	• Developer's Budget					
	• Total cost					
	• Functional Unit (student, bed, seat, room)					
		• Superficial (gross floor area)				
		• Functional Area/Cost				
			• Elemental (group)			
				• Elemental (individual and sub-elemental)		
			• Approximate Quantities (to suit level of detail available)			

Figure 8.2 Cost Planning Techniques and the RIBA Plan of Work

In Australia the equivalent elements and units of measurement are given in the AIQS (2000) Appendix A of that document.

If the information permits, then elements should be measured in accordance with the element unit quantity (EUQ) and priced with suitable rates from previous cost analyses (of similar specification) or other sources of cost information. During the early stages of a project these estimated element costs may be based on the barest of information and may be subject to high levels of inaccuracy. However, as the design progresses these element costs should be reviewed regularly and as more detailed information becomes available more accurate measurements may be made of the quantity and quality (specification) of each element, and eventually down to more detail at the level of each component part of the element, the *sub-element*. Preliminary costs based on derived rates from previous cost analyses or other sources, may give way to ratio and area adjustments, which will eventually be superseded by approximate quantities once suitable drawings have been prepared.

Table 8.2 List of RICS/BCIS Elements and Units of Measurement

Element	Element Unit	Measurement
1. *Substructure*		
1. A substructure	m^2	Area of lowest floor measured as for gross internal floor area.
2. *Superstructure*		
2A Frame	m^2	Area of floors relating to frame, measured as for gross internal floor area.
2B Upper Floors	m^2	Total area of upper floors.
2C Roof	m^2	Area measured overall roof surfaces.
2D Stairs	m	The total vertical height of each staircase and its width between stringers should be given in metres.
2E External Walls	m^2	Area of external walls measured on outer face (excluding openings).
2F Windows and External Doors	m^2	Total area of windows and external doors measured over frames.
2G Internal Walls and Partitions	m^2	Total area of internal walls and partitions (excluding openings).
2H Internal Doors	m^2	Area of internal doors measured over frames.
3. *Internal finishes*		
3A Wall Finishes	m^2	Total area of wall finishes.
3B Floor Finishes	m^2	Total area of floor finishes.
3C Ceiling Finishes	m^2	Total area of ceiling finishes.
4. *Fittings and Furnishings*		
4A Fittings and Furnishings	None	
5. *Services*		
5A Sanitary Appliances	Number	Fittings to be noted as grouped or dispersed.
5B Services Equipment	Number	
5C Disposal Installations	Number	Fittings to be noted as grouped or dispersed.
5D Water Installations	Number	Cold water, hot water, steam and condensate draw-off points noted.
5E Heat Source	kW	Boiler rating in kW.
5F Space Heating and Air Treatment	Tm^3	Cubic content of treated space (Tm^3).

(*Continued*)

Table 8.2 (*Continued*)

Element	Element Unit	Measurement
5G Ventilating Systems	Tm^3	Cubic content of treated space (Tm^3).
5H Electrical Installations	kW Lux Number	Total electric load in kW. Tabulate illumination levels (Lux) by principal function. Total number of power outlets and lighting points.
5I Gas Installation	Number	Number of draw-off points.
5J Lift and Conveyor Installations	Number	Number of rush period passengers. Number, capacity, speed, stops, doors and height of lifts.
5K Protective Installations	Number Tm^2	Number of sprinkler outlets, control mechanism and areas served.
5L Communications Installations	None	Cost of each installation given separately.
5M Special Installations	None	Cost of each installation given separately.
5N Builder's Work in Connection with Services	None	Cost of builder's work in connection with each of the services elements given separately.
5O Builder's Profit and Attendance on Services	None	Cost of profit and attendance in connection with each of the services elements given separately.
6. *External Works*		
6A Site Works	m^2	Mainly areas of site preparation, surface treatments and site enclosure. Fittings and furniture numbered.
6B Drainage	None	Surface water and foul drainage. Sewage treatment.
6C External Services	None	All utilities and builder's work/profit in connection.
6D Minor Building Work	m^2	GFA of ancillary buildings. Alterations – no quantity
Preliminaries	%	% of remainder of contract sum.
Total Cost	£	Total contract sum less contingencies.

In each elemental costing and review the sum of the allocated elemental costs is added together to give the total estimated cost. Within each element the costs are expressed in:

- total element costs;
- cost per square metre of gross floor area (GFA);

- element unit rate (EUR) (based on total element cost divided by the element quantity).

The EUQ for the total element has to be given as noted in Table 8.2 and this allows the EUR to be derived. The standard format for presenting this cost and other data is given in the *Standard Form of Cost Analysis* (RICS, 1969) and an extract of this format is shown in Table 8.3. Also see the Appendix for details of the *Standard Form of Cost Analysis*.

Table 8.3 Format of Cost Analysis

A	B	C	D	E	F
Element and design criteria	Total cost of element £	Cost of element per square metre of GFA £	EUQ	EUR £	Specification
As noted in RICS/BCIS (1969)	The sum of all sub-elemental and other costs	The total cost (B)/GFA	m^2 No Tm^3 kW Lux	Total cost (B)/EUQ (D)	Can be taken to sub-elemental level in many elements

Source: RICS (1969).

Advice on the use of the elemental method in practice
This technique is inherently flexible and can be adapted to suit the level and complexity of information available. The earlier this method is used the greater the potential for gaining an accurate assessment of the costs of the project. Elemental estimating should be used at the earliest stage the project can be reasonably priced, initially using the Group Elements, and eventually translating this into the RICS (1969) Standard List of Elements. The technique can be adapted to suit the various qualities of information existing in each of the elements and it can be taken to sub-elemental level when the project information advances to this stage. This flexible approach has the characteristic of robustness that cost planners admire.

Elemental cost planning requires a commitment to continuous review of the elemental estimate as the design progresses through the various stages. The elemental costs are appraised as new information becomes available and then they are adjusted to take account of specific changes in particular elements. Changes in economic factors, market conditions, prices of building labour and materials generally that affect overall pricing levels throughout the estimate can often be forecast and included in the estimate through the use of building cost indices. A general price adjustment based upon the trend in these indices is important and must be monitored closely by the cost planner. Building cost indices were discussed in Chapters 5 and 6.

Good examples of cost plans taken to the group element level in the Outline Proposals stage are given in Flanagan and Tate (1997: 253). A cost plan to the elemental level of detail at the Detailed Proposals stage (previously the Scheme Design stage) is also given in Flanagan and Tate (1997: 281). For Australian practice at equivalent stages, see Smith and Love (2000: 150–170) together with a Production Information stage cost plan (Tender Documentation).

Advantages of elemental method

- An adaptable system which can be used with suitable adjustment at all stages of the design process.

- Its basic form of analysis (the element) means it is a system capable of being understood by all parties – client, design team members, contractor, subcontractors and users, if necessary!

- Comparisons between projects and between elements can be made rapidly, and each element can be adjusted individually.

- Elemental cost plans allow the design team members to gain elemental costs and to ascertain the cost implications during the early stages of the project.

Disadvantages of elemental method

- Elemental cost plans require more time and effort in their preparation, but this investment should give better returns in aiding decision-making.

- Extensive cost databases are required to ensure accuracy of costs.

- High levels of ability and expertise are required if elements are to be properly evaluated and adjusted.

Approximate Quantities Method

Many cost planners regard this method as the most reliable technique to employ. In contrast to the previous methods this is not a single price estimating method.

Approximate quantities rely upon relatively well-developed design and detailed specification information for their accuracy, and for this reason can only be used in the Detailed Design, Final Proposals and Production Information (or Tender Documentation) stages. As its name suggests it requires information that closely resembles that needed to measure reasonably accurate quantities for an element, or a group of elements. The cost planner uses a modified version of measuring and describing the work and applying rates from the BCIS, Bills of Quantities (BQ) and published price books for pricing. Some cost planning offices have developed their own composite rates that match the modified or simplified form of measurement adopted by the office. The method of measurement and pricing is at the discretion of the user, so systems can vary between offices and individuals. The main features are that they remain flexible, adaptable and accurately represent the cost of work detailed on the drawings.

Method of measurement

The approach to measurement and pricing is similar to that adopted by estimators in pricing 'Builder's Quantities' for projects using only specifications and drawings. Whereas measurement using the RICS and National Federation of Building Trades Employers (NFBTE) (1978) (amended 2000) SMM7 *Standard Method of Measurement of Building Work* requires quantity surveyors to measure in great detail and to define levels of accuracy and description, the approach in approximate quantities is much more informal and flexible. It allows the cost planner to group items together relating them to common forms of measurement, sections of subcontract work or sequences of operations. Major items are grouped together; less significant items such as labours on major items are ignored in measurement, but accounted for by slightly increasing the price of the major item by an appropriate percentage.

An example of costing a ground slab is given using the detail shown in Figure 8.3.

The costs used for this type of analysis can be gained from one or more of the published sources such as *Spons, Laxtons, Rawlinsons* or the *Building*

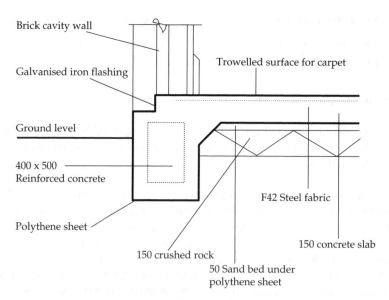

Figure 8.3 Ground Slab and Foundation Detail

Economist. Alternatively, costs may also be obtained from the BCIS cost analyses or internal sources from the cost planner's office for previous projects.

Unit of measurement: Square Metres
Items to be included in 1 m² analysed   £

• Crushed rock/hardcore	1 m²	@	£9.50	9.50
• 50 mm sand bed	1 m²	@	£2.50	2.50
• Polythene sheet (dpm)	1 m²	@	£2.80	2.80
• 150 mm concrete slab	0.15 m³	@	£125.00	18.75
• F42 fabric reinforcement	1 m²	@	£4.25	4.25
• Trowel surface (delete screed)	1 m²	@	£4.00	4.00

Typical cost per square metre   £41.80

Several additional items could be included in the ground slab example given above. For example, it is possible to include excavation items (such as reduced level excavation, disposal and compacting ground) and floor finishes (such as screeds, damp-proof membranes and the finishes). However, this reduces the flexibility of the cost database as it crosses over into too many elements and also includes too many out-of-sequence operations into the one item.

In developing a suitable approximate quantities system the cost planner must balance speed, simplicity and ease of use on one project and all future similar projects. It must be recognised that if, and when, changes in specification do occur, the method adopted must be able to quickly incorporate the changes with minimal disruption.

Approximate quantities are also the most appropriate and accurate for carrying out *cost checks* of individual elements, which occur in the next stages of Tender Documentation, the *Final Proposals* and *Production Information* stages.

Advice on the use of the approximate quantities method in practice

- Subject to the information being available, this is a reliable method of estimating. Quantity surveyors and cost planners are probably most confident with the accuracy and simplicity of this technique.

- It is a method that can clearly reflect any adjustments needed in alternative choices or specifications for quantity for quality.

- Cost information is readily available from various sources; in numerous price books, BQ and build-ups using basic estimating principles. It can also be adapted to suit individual approaches to pricing and measurement.

- However, the greater accuracy of estimating is at the expense of more time being required to complete the estimate. Most would consider this to be a fair trade-off, for example with the use of the cost planner's own and subcontract prices.

- The technique is dependent on suitable cost data being available. For the service installations such as electrical, mechanical, lift, heating and air-conditioning systems, which are often the subject of provisional or prime cost sums in the BQ, it may be difficulty to accurately measure and price using this technique. Thus, in many cases the services installation may have to be priced on the basis of the earlier cost data. Many cost planners and quantity surveyors have recognised this problem and in the last decade greater effort has been made to cost plan, measure and control the costs of services installation, in the same way the design team manages the remaining building elements.

- Naturally, the accuracy of the technique can only reflect the quality of the information provided. The method should not be used prematurely in the design process when too many assumptions have to be built into the technique. Nor should it be applied across all elements during the later design stages if some elements have not been satisfactorily considered. The cost planner should exercise considered judgement as to when this technique should be adopted in an individual element, and it may be the case that a range of techniques is applied to suit the quality and extent of information available in each of the elements.

For practical examples of the approximate quantities method of estimating refer to the further reading at the end of this chapter.

Question 8.1
Budget Distributing Techniques

Describe the two techniques used as a basis for cost planning in the design stages to distribute the costs of a project and identify the RIBA stages in the Plan of Work where they are likely to be used.

OVERVIEW OF DETAILED PROPOSALS ACTIVITIES AND DOCUMENTATION

As in previous chapters the RIBA (2000) Plan of Work provides an excellent overview of all the work stages including the Detailed Proposals stage. The important activities can be summarised as:

- evaluate Outline Proposals to establish compliance with the developing brief;

- ensure in-principle compliance with statutory authorities and advise on cost effects of compliance;

- conduct further cost studies;

- prepare elemental cost plan and update progressively as more information becomes available;

- advise on project programme;

- participate in the completion and signing off of the project (design) brief;

- at end of stage confirm elemental cost plan;

- establish firm cash flow projection; and

- review and update procurement advice taking account of more detailed project conditions.

The basic priorities of this stage are establishing realistic elemental costs that remain within the group elemental (or individual element) targets established in the Outline Proposals stage and to fully reflect the design decisions made to date and during this stage. The information the cost planner has to work with to prepare an elemental cost plan during this stage consists of the following:

- complete project (or design) brief;

- outline cost plan in group elemental format (from previous stage);

- final and developed dimensioned sketch plans;

- elevations and sections of most parts of the project;

- structural sketches with dimensions of all main members indicated;

- preliminary schedule of finishes;

- site layout with disposition and site coverage of building(s), site treatment, location of utilities and landscaping; and

- specification notes for critical elements.

The engineering costs of the project are likely to form a significant proportion of the total cost of the project and by this stage the costs of the various services should now be set out in elemental format and not as lump sums. These elemental services costs should be obtained from the respective engineering disciplines.

The cost output from this stage will be the first detailed cost plan in an elemental cost format. The elemental costs will be the current date project costs plus an allowance for the escalation or inflation costs to the projected tender date. It may also be necessary to identify other costs such a professional fees, loose furniture and equipment and any special supervision or administration costs (project management) as required. These should be separately identified. Whether these costs are included in the cost plan generally depends on the instructions of the client. It is important for the design team to know from the outset whether the budget or cost limit includes these allowances. A substantial amount of money will be needed to cover these in most projects.

If budget or cost limit has been exceeded the cost plan should be accompanied by recommendations on potential savings and changes if necessary.

In terms of cost planning or estimating methods used at this stage, ideally the information should be detailed enough for the cost planner to begin using the approximate quantities method for each element that has reasonably developed data and information. The main methods of estimating the cost of each element are simple proportion, inspection and wherever possible when the information is suitable, approximate quantities.

The cost planner should provide alternative estimates based on cost research for different construction methods, materials, engineering systems, energy saving design (passive measures) features and engineering approaches (active approaches) features during the various activities in this stage. Where required by the client organisation, it is important to prepare and document the cost plan and budget document in accordance with relevant budgeting procedures.

The Detailed Proposals cost plan should be presented on at least an elemental basis with specific construction, finishes and services stated, with elemental unit rates and quantities (if these can be measured) to be shown.

We can use cost analyses of previous similar projects from a suitable cost database such as the BCIS. The BCIS has hundreds of cost analyses that users can gain access to for cost planning new projects. In the same way as described in previous chapters we can interrogate the BCIS to narrow down our focus on projects with similar characteristics to the new project. Our first filtering of the data would note the function, type of work (new build or renovation) to bring us to the group of analysed projects that will form the basis of our refined search. At this stage of our cost planning detail we need the reasonably detailed information provided by cost analyses of similar projects. The format closely follows the RICS (1969) *Standard Form of Cost Analysis* provided in the Appendix. The information contained in a standard cost analysis from the BCIS is shown below in three parts:

1. background project information (Table 8.4);
2. project details (Table 8.5);

Table 8.4 Background Project Information

BCIS Online – Elemental Analysis Number XXXX	A – 2 – 0000
Building Function:	320 – Offices
Type of Work:	New build
GFA:	1,384 m²
Building Function:	320 – Offices
Type of Work:	New build
Job Title:	Offices, No way, Anonymous business park
Location:	Bury St. Edmunds, Suffolk
District:	Deleted
Grid Reference:	Deleted
Dates:	Receipt: 21 August 2000
	Base: 21 August 2000
	Acceptance: 15 September 2000
	Possession: 18 September 2000

Source: Adapted from *BCIS Online*. Details of location and BCIS references suppressed for confidentiality reasons.

3. physical information (Table 8.6);

4. elemental cost data (Table 8.7);

5. project specification (Table 8.8).

Whilst we have broken the cost analysis into its component parts to explain the contents of each part of the analysis, the BCIS data is provided in one whole analysis without this breakdown.

An extract from a BCIS cost analysis is shown in Table 8.4. This first part identifies the function of the analysed building, the type of construction work, physical dimensions in terms of GFA, with sufficient location information to enable the project to be traced and the all important timing of the contract to set the indices for the new project.

The next part of the analysis begins to expand the project information to provide basic data on the general description of the project, site and market conditions, client details and the type of contract including the contract amount and the first breakdown of this total sum into the broad component costs. These are shown in Table 8.5.

A more detailed description of the project is given as shown in Table 8.6 with the essential areas making up the project including GFA, area on each floor,

Table 8.5 Project Details

Project Details:	2-storey office block together with external works including precast and in situ concrete and macadam paving, fencing, landscaping, services, drainage and demolition of existing buildings.
Site Conditions:	Level demolition site with moderate ground conditions. Excavation above water table. Unrestricted working space but restricted access.
Market Conditions:	Negotiated contract. £52,000 fees for architect, structural engineer, planning supervisor, planning/building regulations and services engineer.
Client:	Lotsa Land Ltd.
Tender Documentation:	Schedule of works
Selection of Contractor:	Design and build – negotiated
No. of Tenders:	Issued: 1, Received: 1
Contract:	JCT with Contractor's Design 1998
Contract Period (months):	Stipulated: – Offered: 7, Agreed: 7
Cost Fluctuations:	Fixed
Tender Amended:	Contract sum excludes fees and charges
Tender List:	£2,016,204
	Contract breakdown
Measured Work	£1,570,511
Provisional Sums	£158,951
Prime Cost Sums	£0
Preliminaries	£128,086
Contingencies	£77,160

Source: Adapted from *BCIS Online*. Details of location and BCIS references suppressed for confidentiality reasons.

wall to floor ratio, usable area, number of storeys and functional units where they are appropriate. This part of the analysis expands on the physical attributes of the analysed project.

The fourth section of a standard BCIS cost analysis expands the cost data from total contract sum and broad cost divisions of the earlier part of the analysis to a more detailed cost analysis with each element cost detailed and described in terms of total cost, cost per square metre GFA, EUQ (see Table 8.7), EUR and the percentage that the element cost forms of the total cost of the project. This data is essential for cost planning during the

Table 8.6 Physical Information

Accommodation and Design Features: 2-storey office block. Concrete foundations, RC bed, PCC upper floor; stairs. Steel frame with feature truss supports. Aluminium roof cladding. Facing brick/block walls; Moduclad. Aluminium windows, louvred doors, roller shutters. Brick, block, metal stud and sliding/folding partitions. Flush fire doors. Plaster, emulsion, varnish and tiles to walls; vinyl, carpet and raised flooring; suspended ceilings. Bar/kitchen fittings, vanity units. Sanitaryware. Gas LPHW central heating; comfort cooling, ventilation and electric light and power. Lift. Alarms, data cabling.				
Basement	$0\,m^2$			Area of external walls
Ground Floor	$692\,m^2$			Wall to floor ratio
Upper Floors	$692\,m^2$			Average storey heights
		Basement		$-\,m$
GFA	$1,384\,m^2$	Ground		$-\,m$
		Upper		$-\,m$
Usable Area	$-\,m^2$			
Circulation Area	$-\,m^2$	Internal cube		$-\,m^3$
Ancillary Area	$-\,m^2$			
Internal Divisions	$-\,m^2$	Spaces not enclosed		$-\,m^2$
GFA	$1,384\,m^2$	Number of units		1
Functional Unit	Rate	Storeys as % of GFA		
$1,219\,m^2$ Usable Floor Area	£1,219	2 storeys		100.00%

Source: Adapted from *BCIS Online*.

Detailed Proposals stage, as we shall see when we begin preparing our cost plan.

Another important feature of the *BCIS Online* service is the ability to be able to automatically index and 'rebase' the analysed project to a projected date where the index can forecast the updated index value. In addition the BCIS service can also adjust for the appropriate and selected location index from the original project to the new location. An example of these adjustments is given in the selected project given in Table 8.7.

The final part of the analysed data consists of detailed specification information for each element. This information is sufficiently detailed for a cost planner to be able to identify the element unit costs for that particular specification in the analysed project. So, for instance using Tables 8.7 and 8.8 we can ascertain that a steel frame supporting an area of $1,384\,m^2$ for a new

Table 8.7 Elemental Cost Data

Analysis					
Figures in *italics* rebased to a Tender Price Index of 225 (1st Quarter 2006 (Forecast: 225)) and adjusted using a location index of 110 (Berkshire (Index: 110, Sample: 133)).					
Element	Total Cost	Cost per Square Metre	EUQ	EUR	%
1 Substructure	£72,106	£52.10	692 m^2	£104.20	3
2A Frame (includes others)	£120,580	£87.12	1,384 m^2	£87.12	6
2B Upper Floors	£39,815	£28.77	692 m^2	£57.54	2
2C Roof	Included in element 2A				
2D Stairs	£54,012	£39.03			2
2E External Walls	£127,557	£92.17			6
2F Windows and External Doors	£151,281	£109.31	1,100 m^2	137.53	7
2G Internal Walls and Partitions	£74,523	£53.85			3
2H Internal Doors	£34,900	£25.22	54 No	£646.29	1
2 Superstructure	£602,668	£435.45			31
3A Wall Finishes	£27,798	£20.09	2,986 m^2	£9.31	1
3B Floor Finishes	£77,909	£56.29	1,244 m^2	£62.63	4
3C Ceiling Finishes	£24,065	£17.39	662 m^2	£36.35	1
3 Internal Finishes	£129,772	£93.77			6
4 Fittings	£26,505	£19.15			1
5A Sanitary Appliances	Included in element 5F				
5B Services Equipment					
5C Disposal Installations	Included in element 5F				
5D Water Installations					
5E Heat Source					
5F Space Heating and Air Treatment	£255,201	£184.39			13
5G Ventilating Systems	Included in element 5F				
5H Electrical Installations	£262,654	£189.78			13
5I Gas Installations					
5J Lift and Conveyor Installations	£30,864	£22.30			1

(Continued)

Table 8.7 (*Continued*)

Element	Total Cost	Cost per Square Metre	EUQ	EUR	%
5K Protective Installations					
5L Communications Installations					
5M Special Installations					
5N Builder's Work in Connection	£3,086	£2.23			
5O Builder's Profit and Attendance					
5 Services	£551,806	£398.70			28
Building Sub-total	£1,382,856	£999.17			71
6A Site Works	£230,864	£166.81			11
6B Drainage	£69,444	£50.18			3
6C External Services	£46,296	£33.45			2
6D Minor Building Works					
6 External Works	£346,605	£250.44			17
7 Preliminaries	£128,086	£92.55			6
Total (less Contingencies)	£1,857,548	£1,342.16			96
8 Contingencies	£77,160	£55.75			3
Contract Sum	£1,934,708	£1,397.91			100

Source: Adapted from *BCIS Online*.

build 2-storey office building of this kind can be costed at a rate of £87.12 per m^2 of supported area and represents in the region of 6% of the total cost of the building. Each element that has EUQ provided can be analysed in a similar manner.

Question 8.2
Cost Analyses

- Why are cost analyses important in cost planning?
- Where does the cost data used for the cost analyses come from?
- How are preliminaries treated in a cost analysis or in the published cost books?

Table 8.8 Project Specification

	Specification	
	Element	**Specification**
1	Substructure	32 No foundations pits. C10P concrete blinding, GEN 3 concrete foundations, RC40 beds.
2A	Frame	Structural steel frame, paint and sprayed fire protection; feature external truss supports.
2B	Upper floors	PCC beam and block.
2C	Roof	Aluminium standing seam cladding to roof, Moduclad to walls. Rainwater goods.
2D	Stairs	2 No staircases.
2E	External walls	102 mm facing brick, cavity and 140 mm block inner skin (154 m^2). Circular ornamental bands. Wall cladding included with roof.
2F	Windows and external doors	Aluminium windows with grey tinted glazing to stairwells; entrance and louvred doors. Roller shutter doors. Window boards.
2G	Internal walls and partitions	42 m^2 common brick, 96 m^2 facing brick, 207 m^2 block. PCC lintels. 678 m^2 metal stud and ply/plasterboard partitions; 3 No sliding folding partitions, cubicles.
2H	Internal doors	58 No flush half hour fire doors, ironmongery.
3A	Wall finishes	431 m^2 coat plaster to blocks, 969 m^2 skim coat to studwork. 1,313 m^2 emulsion to walls, 146 m^2 varnish to ply stud; 127 m^2 ceramic tiles.
3B	Floor finishes	1,244 m^2 raised access floor; 803 m^2 vinyl and carpet. Skirting and decoration. Aluminium matwells.
3C	Ceiling finishes	705 m^2 suspended ceilings.
4	Fittings	Kitchen fittings; reception counter, auditorium timber steps, ladder, mirrors, vanity units. Provisional sum (£10,000) for bar fittings.
5A	Sanitary appliances	Sanitaryware, and disabled WCs.
5B	Services equipment	PC sum £10,000 for bar equipment. Kitchen equipment.
5D	Water installations	Water services.
5F	Space heating and air treatment	Gas LTHW central heating, comfort cooling.
5G	Ventilating systems	Extract ventilation with duct pipes.
5H	Electrical installations	Switch gear, electric power and lighting. Fire alarm, telephone and data cabling; lightning protection.

(Continued)

Table 8.8 (*Continued*)

Specification		
	Element	**Specification**
5J	Lift and conveyor installations	Provisional sum £20,000 for lift.
5N	Builder's work in connection	General builder's work.
6A	Site works	Tarmacadam roads and footpaths, thermoplastic markings, 254 × 125 HB kerbs, 50 × 150 PCC edgings. 100 mm dry lean concrete, 60 and 80 mm block paviours. Bollards. Tree pits. Perimeter fencing. Landscaping.
6B	Drainage	Provisional Sum £45,000.
6C	External Services	Water, gas and electricity mains.
7	Preliminaries	7.41% of remainder of Contract Sum (excluding contingencies).
8	Contingencies	4.46% of remainder of Contract Sum (excluding preliminaries).

Source: Adapted from *BCIS Online*.

The Approach to the Detailed Proposals Work Stage D

The order in which elements are considered will usually be dictated by the order in which the plans, specifications and Detailed Proposals are produced. Sufficient information must be produced by the design team and made available to the cost planner before the cost plan is started.

An important skill in the process of developing elemental cost targets for each element in the Detailed Proposals stage is to identify the major differences between the selected rate/entry in the cost analysis and the proposed building. The brief of the new project and all the information on the selected project should be studied and compared. The detailed cost analysis or several analyses should be carefully scrutinised, and then all the expected or defined element specifications of the proposed project should be identified where the information can be gained from the design team.

Where a published source of cost information is being used to gain a suitable element rate the information provided in the publication should be carefully studied and differences noted.

ADJUSTMENTS FOR QUANTITY, QUALITY AND PRICE [Q, Q, P]

The major differences that are likely to come to light in this process can be categorised under three headings:

1. Quantity [Q]
   - A difference in the floor area.
   - Items in the analysed/published project, which will not be required in the new project.
   - Items that are missing from the analysed building, but which will probably be needed in the proposed building.
   - Significantly different requirements in the preliminaries and contingencies elements, for example, temporary access roads, difficult access and storage.

2. Quality [Q]
   - Difficult to identify with the information usually available at this early stage. The statement of quality in an initial brief is usually vague. So vague, in fact, that the relatively small differences in quality which remain after choosing the cost analysis/cost data cannot be allowed for until the next stage, Final Proposals or even later.

3. Price [P]
   - A difference between the date of tender/or publication of the analysed/published building and the date of this first estimate (price level).

Each elemental cost target is prepared by isolating the differences between the effect of each of these factors on the element (as it was designed in the previous building from the cost analyses) and as it will look and designed in the new project. Note that these adjustments are not the *design contingencies*, or *design risk* as these are considered in a separate percentage or adjustment for the whole design.

The differences in Quantity and Quality will vary from element to element, but differences in *price level* are most likely to affect all elements equally from the date of the cost data to the present date. Due to this, it is worth considering how allowances are made for price level.

Adjustments for Price Level

The main adjustments for price level are for:

- differences in general market price level;

- variations between the contractor's price level and general market price level at tender date;

- differences in site conditions and location; and
- differences in contract conditions, weather conditions and the like.

When we use the *BCIS Online* system the adjustment for the new dates is relatively easy to carry out. This process was described in the last chapter, but it is worth repeating why and how the changes for differences in the key dates for the new project are made. The first adjustment noted above is for differences in the general market price level between the date of tender of the analysed building and the receipt of tender of the new project.

The most difficult allowance to make is for a change in general market price level between the date of tender of the analysed building and the date of tender of the new project. It is often not possible to accurately allow for these changes using the relevant values of a price level index. The BCIS All-In Tender Price Index (BCIS TPI) attempts to do this for the cost planner. Naturally, the index is accurate for the differences in price between the date of the analysed project and the last firm index calculation. However, in the forecasts of projected index values there is no guarantee of accuracy and the farther the index is projected the greater the possibility of error. In addition, at any time and for a specific location individual or unique factors may apply to disturb the predicted values. This is where the cost planner has to employ the local knowledge and judgment to adjust the normative values expressed in the index.

When we are preparing the elemental cost targets in the Detailed Proposals cost plan the amounts can be indexed to the present date of when the cost plan is prepared or based on the projected tender date. This process being relatively simple using the rebasing facility of the *BCIS Online* system.

The third adjustment to general price levels is from the projected tender date to the project completion date (escalation date to completion date). The three stages of price adjustment are shown diagrammatically in Figure 8.4.

Identify	STAGE 1	Forecast	STAGE 2	Forecast	STAGE 3	Forecast
SOURCE DATA ACTUAL INDEX	% Escalation to current date	PRESENT DATE INDEX	% Escalation to tender date	**INDEX FOR TENDER DATE**	% Escalation to completion	**INDEX FOR COMPLETION DATE**
1. Date of cost analysis (Source)	⇨	2. Current or Present Date	⇨	3. Expected Tender date	⇨	4. Completion of contract

Figure 8.4 Updating Costs

This is an allowance for price changes that may occur during the contract duration an important consideration is whether the contract is 'fixed price' or subject to fluctuations or 'rise and fall'. This allowance should be included at the end of the cost plan under the heading of fixed price contract or fluctuations. In Australia, this is referred to as *Escalation during Contract Period*. The cost planner must make a genuine pre-estimate of how much the successful tenderer will allow for changes in prices between the date of the tender and completion of the works. In a fixed price contract this allowance is made competitively with the other tenderers, whilst in a fluctuations based contract reasonable reimbursement will be made to the contractor based on the terms of the contract. The full amount of price increases may not always be reimbursed and so the amount may be difficult to predict.

Other allowances

The 'design risk' or 'design contingency' allowance is usually considered as a percentage rather than a lump sum. This was also discussed in Chapter 7 and as we proceed through the stages the percentage and the amount allocated should reduce to reflect the firm design decisions are made.

The *contract contingency* accounts for the unforeseen additional costs that are likely to arise during the development of any project. Items such as variations in site conditions and contract conditions that are obvious at this stage are allowed for in 'preliminaries' and should not be included in this 'contract contingencies' allowance. Similarly, this percentage should be reduced to reflect the smaller element of the *unknown* as we proceed, until we reach the final stage, Tender Documentation, when it should be based on the actual contingencies amount or percentage to be allowed in the contract documents.

Allowances for a project contingency are not usually based on indices, but the cost planner has to assess the additional amount that may be required to cover delays, inflation, fee negotiations and major changes that may be required by the client or other authorities. Again, this allowance is made at the end of the cost plan.

PREPARING COST TARGETS

When a cost target is being prepared for each element, *all* the factors affecting element costs should be considered by the cost planner. The cost planner can only be expected to deal with the actual information generated by the design team and cannot be expected to make assumptions to cover every aspect of the evolving design. Experience and sound lines of communication are required during the development of the project during this stage.

As noted earlier there are three primary factors affecting element cost and these are:

- Quantity [Q]
- Quality [Q]
- Price [P] Level.

All three factors are allowed for in each of the *methods* used to prepare cost targets in the Detailed Proposals stage.

These methods are:

- simple proportion
- inspection
- approximate quantities.

Simple proportion is a useful method to use when the source cost data is a cost analysis or a series of cost analyses of the same building type. It automatically includes allowances for everything required in the element in the particular type of building considered.

When using *approximate quantities* it is easy to overlook details that can only normally be considered and measured during the Final Proposals and Production Information (Tender Documentation) stages. Nevertheless, this method sometimes has to be used, especially when a new requirement for the project is under consideration and can be reasonably defined, so that some reasonable attempt at measurement can be made.

Inspection is usually reserved for elements or sub-elements such as preliminaries, hardware or drainage, for which no effective measure of Quantity has been devised.

We shall now consider each of these techniques in detail.

Simple Proportion

This technique is based on the assumption that the cost of an element is directly proportional to both the general market price level and the Quantity of the element.

The preparation of a cost target occurs in three stages:

- Adjust for Quantity
- Adjust for Price Level
- Adjust for Quality.

Adjust for Quantity
Assuming that the cost of an element is directly proportional to its Quantity

$$\frac{\text{Cost of an element in new project}}{\text{Quantity of element in new project}} = \frac{\text{Cost of element in building analysed}}{\text{Quantity of element in building analysed}}$$

That is, the EUR in new project = EUR in analysis.

The EUR is the rate obtained by dividing the quantity of an element into the total cost of that element. The standard list of elemental units of measurement is defined by the BCIS (1969), AIQS (2000).

Assuming no adjustments for Price and Quality were necessary, the cost target of an element is calculated

$$\frac{\text{Cost (target)}}{\text{Quantity}} = \text{EUR}$$

Therefore, Cost = EUR × Quantity

The adjustment for Quantity is done simply by multiplying the appropriate EUR by the Quantity of the element in the new project.

Simple proportion *cannot* be used when:

- there is no effective measure of Quantity for the element (i.e., no EUR exists); and

- it is not yet known what Quantity of the element will be required in the new project.

Adjust for Price Level
Using the simple proportion method and assuming that the cost of an element is proportional to the price index, then

$$\frac{\text{Current element cost}}{\text{current index}} = \frac{\text{element cost in analysis}}{\text{index at analysis tender date}}$$

Therefore, ratio of costs = ratio of indices
Then, current cost = ratio of indices × cost in analysis.

Example of simple proportion
Where no adjustments are needed for quantity and quality
Cost of element in analysis = £13,600
BCIS TPI at analysis tender date = 131 (3rd Quarter 1995) firm
BCIS TPI at cost plan date = 224 (3rd Quarter 2005) firm

If no adjustments are needed for Quantity and Quality the cost target would be

$$£13,660 \times \frac{224}{130} = £23,537$$

Let us now consider an example of simple proportion in which both Price and Quantity adjustments are necessary.

Where both price and quantity adjustments are needed. No quality adjustment needed.

The following information is for element 2B: Upper Floors from the cost analysis of the offices in Table 8.5, elemental cost data.

- EUR in analysis $\qquad$ = £57.54 per m^2
- Quantity in new project $\qquad$ = 692 m^2
- Total cost of element $\qquad$ = £39,815
- BCIS TPI at tender date of analysed building (4th Quarter 2002) $\qquad$ = 190 (firm)
- Current BCIS TPI (4th Quarter 2007) $\qquad$ = 246 (forecast)

No adjustment for Quality is required.

$$\text{Target cost for 2c: Roof} = £57.54 \times 692 \text{ m}^2 \times \frac{246}{190}$$
$$= £51,553$$

Although these two adjustments can obviously be done at the same time, as above, it is usually more convenient to do them in two stages.

- *Adjust for Price*

  Current element unit rate $\quad = \quad £57.54 \times \dfrac{246}{190}$

  $\qquad\qquad\qquad\qquad\qquad = \quad$ £74.50 per m^2

- *Adjust for Quantity*

  Cost target $\qquad\qquad\quad = \quad$ 692 m$^2 \times$ £74.50

  $\qquad\qquad\qquad\qquad\qquad = \quad$ £51,554*

*Slight difference due to rounding off

Adjustment for price as the first step is preferred because in making the calculation of the current EUR this figure could be used in other cost plans and provides current cost data that the cost planner will find helpful.

While preparing a cost plan, the cost planner will probably seek reassurance that the cost allowances being adopted are reasonable. The obvious way to do this is to compare the cost targets with the costs of the corresponding elements in other buildings.

For instance, comparing the total elemental cost figures like £39,815 and £51,553 for Upper Floors will not tell you very much; you should compare the Roof's *element costs per square metre* or *EUR* and these unit rates provide the cost planner with a yardstick or benchmark for evaluating the values. In other elements where they may have different EUQ, EUR or more appropriate and consistent measures may have to be used. Also note that a number of elements are not measured in square metres: 5A Sanitary Appliances and 5F Space Heating and Air Treatment, for example.

All cost targets are entered on the cost plan elemental summary using the following headings:

- element
- total cost of element
- costs per square metre GFA ($£/m^2$)
- EUQ
- EUR.

These headings are taken from the detailed cost analysis data form of the RICS (1969). The specification and design notes can be documented using the amplified analysis forms from the same source.

Making adjustment for Quality
This is usually the least straightforward of the three adjustments. The reason for this is that the Quality of each element cannot be specified explicitly at all, and may not be specified expressly at all, and may not be specified fully until the Production Information stage F. What must be done at this stage is to make sure that the cost target for each element is neither too generous, nor too small for the general quality of the materials that the design team intends to adopt during the Production Information stage. It might seem difficult for the design team to determine the cost of the 'general quality' that they intend to adopt before they consider the details of specification for each element. However, members of the design team do not have to start from 'scratch' when they try to fix a cost to their vague notions of the quality that they intend to specify in the later Production Information stage.

By virtue of the fact that they have used a cost analysis rate from a similar project or from the BCIS, means they have, at least, included this specification in their cost plan. Whilst the full design and specification may not be known the inclusion of such a cost should provide a reasonable basis for cost planning. By examining the specification of the element in the cost analysis, cost database or building price publication, the design team can usually detect any parts of the specification that is of a different quality from what they envisage using in the new project.

That is, it is easier to spot differences and make realistic allowances for them than it would be to fix a cost to partially formed ideas about the new project, on their own.

The adjustments made for a change of the Quality of an element are often expressed as percentage additions to (or deductions from) the costs obtained after adjusting for Price and Quantity.

Example of quality adjustment
Element 2F: Windows and External Doors
The existing building analysed had clear anodised aluminium framing and clear float single glazing in fixed (50%) and sliding (50%) windows. The design team decide they would like to have better quality windows in the new project. Bearing in mind their cost limit and the requirements of the other elements, the design team decide to increase the allowance for the element by 20% and to explore further during the Final Proposals and Production Information stages exactly what kind of windows they can afford.

The figure obtained from the cost analysis after adjusting for Price and Quantity is £75,860.

The cost target is adjusted for quality: £75,860 + 20% = £91,032

Rather than decide by inspection on a percentage increase or decrease to allow for a change in quality, *approximate quantities* are often used for this step in *simple proportion*. That is, the detailed design of the analysed building is altered for that part of the specification of an element in the cost analysis that does not seem to be of a suitable quality for the new project. Approximate quantities are measured from this design, and priced in the traditional way. The change in cost is then expressed as a percentage of the total element cost in the analysis and this percentage is added to (or deducted from) the figure obtained by adjusting for Price and Quantity.

Note that using this method does not commit the design team to finally using this specification for the part of the element concerned, but simply to using one of a similar general quality, and as a consequence, one of a similar cost. This allows the design team some freedom of choice, but within constraints.

Although the approximate quantities method has to be used for certain types of quality adjustment, the simple proportion method still retains its advantage over the inspection and approximate quantities methods of preparing complete cost targets.

Basic advantage of the simple proportion method
A simple proportionate adjustment to an element cost in a cost analysis automatically includes allowances for everything needed for the element in the particular type of building considered.

Question 8.3
Simple Proportion

Carry out a *simple proportion* adjustment to the cost of the external walls element (2F Windows and External Doors) from the cost analysis given in Tables 8.2–8.5 and calculate the EUR with the following details:

Cost analysis data
- Specification: Aluminium windows with grey tinted glazing to stairwells; entrance and louvred doors; Roller shutter doors; Window boards.

- Date of tender   = 21 August 2000

- EUR in analysis   = £137.53

- GFA in analysis   = 1,384 m^2

- EUQ in analysis   = 1,100 m^2

New project data
- Similar specification allowed for in new project. No adjustment for Quality is necessary.

- Total floor area in new project   = 2,216 m^2

- EUQ required in new project   = 946 m^2

- Date of cost plan   = October 2006

Summary of simple proportion method
The simple proportion method

- Adjust for Price: the EUR from the cost analysis is updated.

- Adjust for Quantity: the current EUR is multiplied by the
                                  Quantity of the element that will be
                                  required in the new project.

                                  This figure is then reduced to cost per
                                  square metre by dividing it by the total
                                  GFA of the proposed building.

- Adjust for Quality: differences in either the quality of
                                  materials or the relative proportions of
                                  the various qualities of specification
                                  within the element are allowed for by
                                  adjusting the two figures above.

- The total target cost and the target cost per square metre of GFA are
  entered in the Cost Plan Summary Form (RICS, 1969).

The Price and Quantity adjustments are usually done before the Quality
adjustment because:

- They *always* have to be made, and tend to have more effect on the total
  cost figure than the Quality adjustment, and

- The Quality of an element is more likely to change later in the design
  process than the Quantity of the element.

The total cost figure is obtained by adjusting for Price and Quantity and then
reduced to a cost per square metre of GFA. This is done even in the case
when an adjustment for Quality is likely to follow.

The reasons for this are two-fold:

- Comparing the current cost per square metre of GFA with the costs of
  the same element in other buildings is one way of assessing the quality
  of the element in the building analysed, and, less importantly,

- Cost planners like to reduce all costs to costs per square metre of GFA.
  This is because it is a consistent measure and cost planners carry costs
  per square metre of different types of building, almost as their *stock in
  trade*. Many publications, as noted in earlier chapters, list costs of
  various building types under costs per square metre of GFA.

Inspection

This consists of assembling a range of costs of corresponding elements from
previous projects to suggest a pattern from which suitable cost targets can be
chosen.

Example of the use of the inspection method
Element 2H: Internal Doors. This element, which is seldom of great cost significance in terms of the total cost, is one where the inspection technique may be used. (Approximate quantities can also be used if the cost as a proportion of the total cost justifies the time required for measurement.)

The costs of Internal Doors in four analysed office buildings are as follows in Table 8.9.

Table 8.9 Analysis of Internal Doors in Four projects

Building	Tender Date	Preliminaries	Cost of Internal Doors £ per Square Metre/GFA
Office A	January 2002	8.21%	£29.50
Office B	March 2003	11.01%	£28.00
Office C	February 2001	6.78%	£31.50
Office D	September 2000	10.61%	£22.40

The preliminaries element (7) is also shown because some of the elemental costs may be increased to 'load' some of the preliminaries costs into the elements that will be completed earlier in the construction programme. This is a pricing strategy used by some contractors. We need to see the preliminaries cost percentage to judge whether this has happened and in such cases we would be unlikely to use these analyses as a basis for our new project. In Office C and possibly Office A the preliminaries percentages are substantially lower than the other two analysed projects and should be treated with caution, or an adjustment made to the elemental costs presented. The preliminaries percentages in Offices B and D look very similar and closer to the level we would expect in a new project.

The average cost based on the rates for Offices B and D is £28.00 + £22.40 = £50.40/2 = £25.20. Round up to next whole amount = £26.00 per m^2 GFA.

Decision:
After making approximate adjustments for the differences in the preliminaries percentages, we decide to allow £29.00 per m^2 for Internal Doors. That is, an additional +£3.00 per square metre GFA. If the GFA of the new project was 1,440 m^2 the total cost of this element would be

Total element cost $= £29.00 \times 1,440 \, m^2$

$= £37,440$

Question 8.4
Cost Targets by Inspection

You are cost planning a new 4-storey office block and are considering the cost target allowance for preliminaries and Wall Finishes (3A) based upon information from several analysed projects that is available from the BCIS or building price books such as *Spons, Laxtons* or *Rawlinsons*. An extract and summary of the cost data from typical cost analyses of similar buildings is given below:

Building	Preliminaries % BC	Wall Finishes % BC	Cost of Wall Finishes £ per square metre/GFA
Office Building: Single Storey	8.0	1.3	£9.00
Office Building: 2 Storey	8.0	2.1	£17.00
Office Building: 3 Storey	8.2	2.0	£18.00
Office Building: 4–7 storeys	9.8	1.8	£18.00

The overall Building Cost Target is £850,000. The GFA of the project is estimated at 790 m².

For each element determine a suitable percentage allowance, an overall cost target and cost per square metre based on the GFA.

Using the inspection method
This method is used when measures of quantity are not available and considerable time would be required for measurement of approximate quantities. Another factor is the quality standard of the element in question is not unusual or unique.

Approximate Quantities

In this method (described in the previous chapter as a budget distributing technique), a relatively detailed specification is required (and prepared) for the element. Approximate quantities are then measured from this specification and priced in the traditional way.

Example
Element 3B: Floor Finishes
Instead of a single floor finish to all corridors, entrance and common areas the architect has indicated that the entrance area finishes should be of a high

standard and other circulation areas should be of budget grade. It is advised that floor finishes to the office areas are the tenant's responsibility.

The areas can be easily measured and we decide it is best to use approximate quantities based upon an assumed specification. We assume terrazzo in the entrance areas and medium duty vinyl sheet in the other circulation areas. The GFA of the project is 2,216 m^2.

Entrance areas

	£
• 37 mm thick in-situ terrazzo paving 52 m^2 @ £275.00 (BCIS or building price book)	14,300

• 2.0 mm thick flexible medium duty vinyl sheet
   (p.c. range £14.50–£35.00 m^2/supplied)
   188 m^2 @ £37.50 (building price book),
	7,050

	£21,350
• Allow for skirtings, sundry items say +15%	3,200
Total cost target for element	£24,550

Reduce to cost per square metre GFA

$$= \frac{£24,550}{2,216 \text{ m}^2} = £11.08 \text{ per m}^2 \text{ GFA}$$

- For the EUR the total cost of the element is divided by the EUQ. That is, 52 m^2 of terrazzo paving + 188 m^2 of vinyl = 240 m^2 total EUQ

$$= \frac{£24,550}{240 \text{ m}^2} = £102.30 \text{ per m}^2$$

Summary results

Target Cost for 3B Floor Finishes	Total Element Cost	Cost per Square Metre GFA	Element Quantity	EUR
	£24,550	£11.08	240 m^2	£102.30

Note that this does *not* commit us to finally using this specification, but simply to using one of a similar general quality. At this stage, we are only involved in making a reasonable allowance within which floor finishes of the desired quality can be designed during the subsequent Final Proposals and Production Information stages.

To emphasise this again, in the process of preparing approximate quantities at this stage:

- The specification prepared for this technique is not as detailed as the one prepared during the Final Proposals and Production Information stages, and

- The design team is not committed to finally using the specification prepared for approximate quantities, but simply to using one of a similar general quality.

Question 8.5
Approximate Quantities

Prepare a cost target by approximate quantities for the Internal Doors element (2H) with the following information:

- GFA   790 m²

- Number of doors   20 No

- General Indicative Quality   2040 × 820 hollow core flush door; Pacific Maple veneer; rebated timber frame; high quality lock; paint finish.

Using the BCIS or one of the building price books mentioned earlier, prepare an elemental cost target for the internal doors element. Do not forget to make an allowance for price escalation between the cost analysis base dates (assumed to be the 1st Quarter 2005 with an index 220) or of the cost publication and to the present day (assume an index forecast of 238).

Contract Contingency and Design Risk

Before considering each of the elements individually, consideration must also be given to the allowances made for contract contingency and design contingency. This is referred to in some of the texts as *Price and Design Risk*. Similarly, costs in the analysis should be rebased on the *BCIS Online* system (or carried out manually if not) to take account of escalation or increased costs from the present or date of the analysis to the tender date. These were described in Chapter 6 and provision has to be made for them in the cost plan. After all the elemental costs have been summarised on the summary cost plan form these additional allowances are made. It is worthwhile to reflect on the progress of the documentation, its extent and accuracy, and decision-making generally when considering these allowances.

The design contingency allowance should be smaller than the one made during Outline Proposals. Similarly, the contract contingency will also usually be less at this stage than at the earlier Outline Proposals stage. The reason for this is that the design should have advanced considerably during the Detailed Proposals stage, so the *design risk* part of the allowance can safely be decreased in most cases.

The order in which the elements are considered for estimating their costs will vary from project to project. The one important rule that can be made about this order is that the preliminaries and all the contingencies should be considered after all, or most of the other elements have been priced.

One reason for this is that preliminaries, design and contract contingencies are based upon the whole project and the cost planner can best give consideration to those when the design factors in the elements have been reviewed (see Chapter 3). It is *not* good practice to distribute the preliminaries and contingencies allowances throughout the other elements. Similarly, if cost analyses or cost records contain these items in the elements then an attempt should be made to isolate them and only to include element costs without the addition of preliminaries and contingencies within them to ensure consistency of pricing.

Locality allowances have not been forgotten! Again, these should be added after the elemental costs and other allowances have been included. Locality allowances were discussed in Chapter 7.

All these individual and grouped elemental costs are now transferred to a cost plan elemental summary. A suggested format is provided in Table 8.10. This provides a summary of the elemental costs in the format described earlier following the format of the element cost analysis suggested in BCIS (1969) and shown in the extracts presented earlier in this chapter.

The element quantities and unit rates have been inserted for those elements that have been measured. This form can be typed on any word processing program and saved, but it is best produced on one of the standard computer spreadsheet programs so that the rows and columns of the form are automatically calculated and are always correct! This is an appropriate time at which to start producing this standard form if you have not already done so.

Table 8.10 Detailed Proposals Cost Plan Form

New Factory GFA = 538 m² UK Mean Location	Tender Date = 1Q 2006		All Costs Indexed to Tender Date		Project No: 06/88
Element	Total Cost £	Cost/m² GFA $	Element Quantity	EUR $	% Building Cost
1. Substructure	40,900	76.02	521 m²	78.50	9.61
2. Superstructure					
2A Frame	14,470	26.90	521 m²	27.78	3.40
2B Upper Floors	1,955	3.63	53 m²	36.85	0.46
2C Roof	1,665	3.10	7 m²	238.26	0.39
2D Stairs	90,610	168.42	510 m²	177.67	21.29
2E External Walls	85,105	158.19	561 m²	151.70	20.00
2F Windows and External Doors	6,780	12.60	12 m²	564.90	1.59
2G Internal Walls and Partitions	9,435	17.54	23 m²	410.28	2.22
2H Internal Doors	11,010	20.47			2.59
2. Superstructure Total	221,030	410.85			51.95
3. Finishes					
3A Wall Finishes	3,145	5.85			0.74
3B Floor Finishes	4,405	5.29	56 m²	50.82	1.04
3C Ceiling Finishes	7,865	14.62	56 m²	140.46	1.85
3. Finishes Total	15,415	25.76			3.62
4. Fittings and Furnishings	4,090	7.60			1.12
5. Services					
5A Sanitary Appliances	9,865	18.34	5 No	1,937.40	2.32
5B Services Equipment					
5C Disposal Installation					
5D Water Installation					
5E Heat Source					
5F Space Heating					
5G Ventilating System	3,145	5.85			0.74
5H Electrical Installation					
5I Gas Installation					
5J Lifts and Conveyors	6,290	11.69			1.48
5K Protective Installation	37,750	70.17			8.87

(Continued)

Table 8.10 (*Continued*)

New Factory GFA = 538 m² UK Mean Location	Tender Date = 1Q 2006		All Costs Indexed to Tender Date		Project No: 06/88
Element	Total Cost £	Cost/m² GFA $	Element Quantity	EUR $	% Building Cost
5L Communications Installation					
5M Special Installations					
5N BWIC with Services					
5O Profit and Attendant Services					
5. Services Total	57,070	106.05			13.41
Building Cost Sub-total	£338,485	£629.15			79.54
6.External Work					
6A Site Works	42,055	78.17			9.88
6B Drainage					
6C External Services					
6D Minor Buildings					
6. External Work Total	42,055	78.17			9.88
Preliminaries	22,020	40.93			10.58
Contingencies					
Project Cost	£425,560	745.38			100.00

Detailed Proposals Cost Plan

		£
Summary		
Project Cost: current date (1Q 2006)		425,560
Design Contingencies	+2.5%	10,640
		£ 436,200
Contract Contingencies	+2.5%	10,910
		£ 447,110
Escalation to Tender	Included above	
		£ 447,110
Escalation During Contract	+2.0%	8,940
Total Project Cost		£ 456,050

Note: Costs rounded off to nearest £10

SUMMARY OF BUDGET AND COST PLANS

Assuming a budget or cost limit of £425,000 the cost plan information from this Detailed Proposals stage and the previous Outline Proposals stage is given in Table 8.11.

Table 8.11 Cost Plan Summary

Stage	Budget	± Expenditure	Comment
Budget	£425,000		
Outline Proposals	£429,675	+£4,675 (+1.1%)	No budget increase
Detailed Proposals	£456,050	+£31,050 (+7.31%)	Savings needed

If the cost limit was exceeded by the amounts shown above then the design team would have to make the necessary savings to bring the project closer to the budget of £425,000. This should not be a difficult task as the amount above the budget only represents an amount of around +7%. Nonetheless, the design team should work actively to reduce the project cost to within the cost limit as noted.

Chapter Review

As we have seen the Detailed Proposals stage is the point where the cost planning activity increases caused by the rapid rate of increase in Production Information. Some elements have quite detailed information; others may have progressed very little from the Brief stage. The determination of elemental cost targets creates the need to have a number of methods of adjusting the elemental cost accurately and realistically using information at hand. A sensitivity to the client and the design team's requirements is an essential feature of the cost planner's activities at this stage. The planning and controlling of the budget has reached a decisive stage and it is essential that the cost planner keeps a firm, but fair, grip on the decisions being made during this stage. The maintenance of the overall cost limit is of paramount importance and the application of the basic cost planning principles of establishing a frame of reference and then checking the cost target must be vigorously sustained by the cost planner. The third part of the cost planning principles, remedial action, that is, the control part of the process, is started during this stage, but is more actively pursued during the next stage when more detailed and accurate information becomes available for a wider range of elements.

This chapter has reviewed the use of three common techniques of adjustment; simple proportion, inspection and approximate quantities. All have particular relevance depending upon the accuracy and extent of the available information. The situation where these techniques are likely to be used has been simulated in this chapter with the type and quality of information likely to be available. We expect that you will be better prepared to carry out the necessary adjustments to particular elements and, what is more important, can establish and sustain realistic cost targets to carry into the next cost planning stages; Final Proposals and Production Information (Tender Documentation). That is, the controlling of costs begins in earnest during these final stages of the cost planning process.

Upon completion of the Detailed Proposals stage the agreed project brief becomes the final statement of project requirements developed in conjunction with the client and the design team. Only the client has the power to modify the brief after this stage.

Summary of the methods of adjustment

1. Simple proportion
   The element cost from a cost analysis or cost record or database that was identified at the Briefing stage is adjusted by proportion for Price, Quantity and Quality.

2. Inspection
   The target cost per square metre of GFA is chosen from a range obtained from a selection of suitable cost analyses and cost studies.

3. Approximate quantities
   A notional specification of the intended quality is prepared for the element and priced by approximate quantities.

Regardless of which technique, or combination of techniques is used to arrive at a cost for an element, the resulting cost target should be regarded as a reasonable allowance within which the element may be designed during the Tender Documentation stages. The design team should never feel obliged to use exactly the same specification upon which the cost target is based.

Each of these methods of preparing cost targets takes into account all the factors that affect element costs.

Factors	*Methods*
• Quantity	• Simple proportion
• Quality	• Inspection
• Price level	• Approximate quantities

When all the cost targets have been entered on the cost plan summary their total should be compared with the *cost limit*, and remedial action taken if necessary.

You may be thinking that the preparation of a cost plan must take a long time; in practice it can take less time than you have spent reading this chapter.

Review Question

As noted in this chapter the factory and office's project has run into problems with meeting the budget of £425,000. The Detailed Proposals cost plan just calculated indicated that a cost overrun of £31,050, or +7.31% of the budget could be expected. The design team now need to focus upon those areas that may provide the most fertile areas for savings in the period between the end of the Detailed Proposals stage and the Final Proposals and Production Information (Tender Documentation) stages. As cost planner for the project you have been requested to review the design and the cost plan and to make recommendations for possible areas of savings. Whilst there will be no opportunity for a rapid re-design to incorporate the changes into this Detailed Proposals stage cost plan, you should identify the elements and sections of the work that should be given the highest priority for design development at the next stage.

Do not include a detailed cost assessment of the changes at this stage. These will be dealt with at the Tender Documentation stage. However, you may indicate the general order of cost savings for each suggested area of savings. You should also answer the basic question of whether the project budget of £425,000 is still feasible.

References

Australian Institute of Quantity Surveyors (AIQS) *Building Economist*, Quarterly Publication, AIQS, Canberra, Australia.

AIQS (2000) *Cost Control Manual*, National Public Works Council Inc., Canberra, Australia.

The Royal Institution of Chartered Surveyors (RICS) (1969, Reprinted 2003) *Standard Form of Cost Analysis*, RICS, London, UK.

BCIS Online (1998) http://www.bcis.co.uk

Laxtons Building Price Book (2006) Johnson, V. B. (editor), Elsevier Press, Oxford, UK.

Raftery, J. (1985) *Technics of Project Cost Appraisal*, Technical Research Centre of Finland, Research Notes 468, Government Printing Centre, Helsinki, Finland.

Rawlinsons (2006) *Rawlinsons Australian Construction Handbook*, Rawlhouse Publications, Perth, Western Australia, Australia. Annual Publication

Royal Institution of British Architects (RIBA) (1998) *Architect's Handbook of Practice Management*, 6th edition, RIBA Publications, London, UK.

Royal Institution of British Architects (RIBA) (2000) *The Architect's Plan of Work*, RIBA Publications, London, UK.

Smith, J. and Love, P. (2000) *Building Cost Planning in Action*, University of New South Wales Press, Sydney, Australia.

Spon's Architect's and Builders' Price Book (2006) Davis Langdon (editor), E & FN Spon, London, UK.

Further Reading

Ferry, D. J. Brandon, P. S. and Ferry, J. D. (1999) *Cost Planning of Buildings*, 7th edition, Blackwell Science, Oxford, UK.
    Chapter 16: Cost planning at scheme development stage, pp. 213–236.

Flanagan, R. and Tate, B. (1997) *Cost Control in Building Design*, Blackwell Science, Oxford, UK.
    Chapter 6: Cost control during Scheme Design, pp. 57–72.
    Chapter 10: Cost planning during feasibility, Outline Proposals and Scheme Design, pp. 189–225.
    Chapter 11: Cost example from feasibility to Scheme Design, pp. 245–282.

Jaggar, D. Ross, A. Smith, J. and Love, P. (2002) *Building Design Cost Management*, Blackwell Publishing, Oxford, UK.
    Chapter 7: Design cost management: sketch plan stage, pp. 77–83.

Seeley, I. H. (1996) *Building Economics*, Macmillan, London, UK.
    Chapter 7: Cost planning theories and techniques, pp. 125–143.
    Chapter 9: Practical applications of cost control techniques, pp. 182–209.

Ashworth, A. (2004) *Cost Studies of Buildings*, 4th edition, Pearson Prentice Hall, Harlow, Essex, UK.
    Chapter 7: Indices and trends, pp. 117–138.
    Chapter 8: Cost analyses, pp. 139–149.

SUGGESTED ANSWERS

Question 8.1: Budget Distributing Techniques

Describe the two techniques used as a basis for cost planning in the design stages to distribute the costs of a project and identify the RIBA stages in the Plan of Work where they are likely to be used.

The methods of estimating for cost planning the RIBA design stages (budget distribution) are described in the chapter and the further reading, but the essence of each method is given below.

As for the RIBA stage at which each method is used the important factor to remember is that a technique can only be used when suitable information to operate that technique becomes available, see Figure 8.2. The design team should produce more detailed project data during these design stages and the cost planner should not continue using the coarse method of estimating (budget setting) any longer. These should have been clearly superseded in application by the more accurate techniques of elemental cost planning and approximate quantities described below. These techniques are designed to use the more detailed information in an appropriate way.

Elemental method
The elemental method can be applied in a flexible manner to suit the level and quantity of design information that has been created by the design team. As the information database expands the technique expands accordingly to accommodate the information.

The method is based on the breakdown of a project into the standard elements (see RICS, 1969 for *Standard List of Elements* and the details are shown in the Appendix) and costing each element, or sub-element, from past projects and published EUR. See Figure 8.2, Flanagan and Tate (1997), Smith and Love (2000) for examples of elemental cost planning if you are still unsure about the details of this method.

Stage used: The elemental method is a technique used in the later stages of the process. It starts in a rudimentary form in stage C Outline Proposals with group elements, develops further and more extensively into the two subsequent stages D Detailed Design, E Final Proposals and F Production Information where more information is created. The elemental cost plan is the basis for all the cost checking and remedial activities during the final design stage F Production Information, see Figure 8.2.

Approximate quantities
This method measures the project in more detail based upon an elemental, sub-elemental, trade or sectional sub-division and applies rates to the all-inclusive quantities. It is a relatively more accurate method of estimating, but it requires more detailed information and takes longer to prepare. Whilst whole projects may be priced on the basis of approximate quantities it is more commonly used in present practice within the elemental method to measure and cost individual elements and sub-elements. Examples of this method are given in this chapter, references and the further readings.

Stage used: Approximate quantities can, and should, be used from stage C Outline Proposals through stages D and E and on to the final design stage of Production Information (stage F). See Figure 8.2.

These two techniques are generally used after Outline Proposals although they may have some application during this early stage. For instance, for measuring the external works element, which generally will have little relationship to the analysis or analyses being used because the external works component elements are site dependent.

Question 8.2: Where does the cost data for cost analyses come from?

Bills of Quantities (BQ) are a major source of cost information to the cost planner, but the information contained in them must be used with great care. Although a range of 4–10 tenders may vary by only up to 10% in total, individual trade sections may differ by as much as 40% and individual items by up to 200%.

The reasons for such variations in rates are many and may be due to the influence of some of the following broad factors:

- type of project;
- size of project;
- regional location/locality;
- contract conditions;
- market conditions.

Added to these broad factors are variations that may be accounted for by the *vagaries of tendering* due to the conditions applying to the specific project, and of critical importance is the estimator's skill, experience and time made available to prepare the tenders. These can be summarised as follows:

- location of site;
- distribution of preliminaries items;

- errors;

- deliberate distortion of rates or trades;

- lack of accurate cost data;

- security of competition;

- sub-contractors' and suppliers' prices;

- site techniques;

- assessment of standards and workmanship;

- availability of labour;

- financial conditions;

- special requirements of project – security, speed and high quality.

How are preliminaries treated in a cost analysis?
The preliminaries section of BQ may present problems in analysis when the estimator/tenderer has included a large proportion (or the whole) of the preliminaries costs throughout the trades. Ideally, the preliminaries section should reflect the *true* cost of those items included there, and likewise with the cost of the trades and the elements. If the preliminaries section does appear to have been adjusted to an unusually low figure then the cost analyst is left in a difficult position. The preliminaries costs cannot be adjusted at this stage, and the best approach is to include a note or comment to the effect that the preliminaries cost (and percentage of the total) appears abnormally low, and some part of their cost may have been distributed in all or some of the elements (trades). A cost planner seeing this note will then be warned about the risk of error in using the cost data from this analysis.

The inclusion of a suitable percentage for preliminaries was discussed in Chapter 7.

Question 8.3: Simple Proportion Adjustment on Element 2F Windows and External Walls

Check index values on BCIS or any other source used by the cost planner.

- BCIS TPI at analysis tender date 3Q 2000   =   162 (firm)

- BCIS TPI at analysis tender date 4Q 2006   =   234 (forecast)

Adjust for Price
EUR in analysis   =   £137.53

Thus, current EUR	=	£137.53 × $\dfrac{234}{162}$ (or 1.444)
	=	£198.65

Adjust for Quantity

Total cost target	=	£198.65 × 946 m^2
	=	£187,923

As an addition to the requirements of the question we will present the cost per square metre of GFA, since this must also be entered in the cost plan.

$$\frac{£187,923}{2,216 \text{ m}^2} = £84.80$$

Result:	Cost Target for External Walls	Total £187,923	Cost per square metre £84.80

The format we will adopt is that used on the BCIS (1969) *Standard Form of Cost Analysis*:

Element	Total Cost	Cost per Square Metre	EUQ	EUR	%
2F External Walls and Windows	£187,923	£84.80	946 m^2	£198.65	x%

Question 8.4: Cost Target Adjustment by Inspection

Preliminaries

A review of the percentages from previous projects indicates a close similarity and the range runs from 8.0% to 8.2% up to 3 storeys and 9.8% for the 4–7 storey buildings. The latter higher percentage reflects the additional preliminaries items required by high-rise buildings generally. Our building just falls within the 4–7 storey category and we suggest that we must increase the preliminaries percentage from the low levels shown for the offices up to 3 storeys, but our building will probably not reach the levels indicated by the full 9.8% shown for this category overall. Thus, we recommend that a percentage of 9.0% be adopted for the preliminaries element.

As you may be aware, some tenderers price some of their preliminaries in the elemental sections of the building, in the cases presented there does not appear to be an obvious 'loading' or 'stripping' of the elemental costs that would require us to make adjustments in the elemental costs.

Wall finishes
The percentages for this element show a remarkable similarity ranging from 1.3% to 2.1%. The two building forms closest to our project are the 3 storeys and the 4–7 storeys, and the percentages are 2.0% and 1.8%, respectively. Playing it safe we are going to recommend a percentage midway between the two values at 1.9%.

It is well to remember that an approach using inspection as the method of assessment is well suited to elements that do not represent a high proportion of the total costs of the project. An error of 50–100% in the assessment will not cause undue problems in the case of the Wall Finishes element. The preliminaries element is an important element, but neither simple proportion nor approximate quantities can be used successfully on this element. The inspection method is the most common form of assessment method used for preliminaries during the Detailed Proposals stage and the following Tender Documentation stages. If any other of the elements were high-cost elements then either simple proportion or approximate quantities should be used.

The calculations for the cost targets are as follows:

Preliminaries

$$\text{Total Building Cost Limit} \quad = \quad £850,000 \times 9\% = £76,500$$

$$\text{Cost per square metre/GFA} \quad = \quad \frac{£76,500}{790\,\text{m}^2} = £96.84 \text{ per m}^2/\text{GFA}$$

Wall finishes

$$\text{Total Building Cost Limit} \quad = \quad £850,000 \times 1.9\% = £16,150$$

$$\text{Cost per square metre/GFA} \quad = \quad \frac{£16,150}{790\,\text{m}^2} = £20.44 \text{ per } £\text{m}^2/\text{GFA}$$

Summary of allowances

Element	Total Cost	Cost per Square Metre	EUQ	EUR	%
2A Wall Finishes	£16,150	£20.44	N/A	N/A	1.9
7 Preliminaries	£76,500	£96.84	N/A	N/A	9.0

Question 8.5: Cost Targets Using Approximate Quantities

You can use one or more of the published price books such as *Spons* or *Laxtons* for pricing this element. In Australia you could use *Rawlinsons* or the

Building Economist could equally be used for this exercise. These internal doors are priced as follows:

			£
Timber doors	Hollow core flush doors		
	2040 × 820 standard hollow core flush door; prime coated hardboard faced including fully welded zinc annealed frame; average quality mortice lock; furniture and paint finish	No.	355.00
ADD for:	Rebated timber frame and architraves		161.00
	Pacific Maple sliced veneer		13.00
	High quality locks		95.00
		£	624.00
ADD	Sundries and undefined items (say)		
		+15%	93.60
	Cost per Door	£	717.60

Adjust for price: BCIS TPI   1Q 2005   =   220
                     BCIS TPI   present date   =   238

$$\text{Doors}\quad £717.60 \times \frac{238}{220} = £776.31 \text{ per m}^2 \text{ (say) } £780.00$$

Adjust for quantity:

Doors 20 No. @ £780.00	£15,600
Total cost of Internal Doors	£15,600

$$\text{Cost per square metre} = \frac{£15,600}{790 \text{ m}^2} = £19.75 \text{ per m}^2/\text{GFA}$$

$$\text{EUR:}\quad \frac{£15,360}{20 \text{ No}} = £780.00 \text{ per door}$$

Adjust for quality:
Adjustment made above in the approximate quantities.

Summary

Element	Total Cost	Cost per Square Metre	EUQ	EUR	%
2H Internal Doors	£15,600	£19.75	20 No	£ 780.00	x%

Final comment

We should repeat that including this specification in the elemental costing implies no final decision. Through this exercise in approximate quantities we have allocated a cost target of £15,600 for the Internal Doors and within this sum the design team have the freedom to make a selection of an appropriate specification. Ideally, the cost target should be adequate for the design team to have a reasonable choice!

Table 8.12 Summary Elemental Costs

Element	Total Cost £	Cost/m² GFA $	Element Quantity	EUR $	% Building Cost
1. Substructure	40,900	76.02	521 m²	78.50	9.61
2A Frame	14,470	26.90	521 m²	27.78	3.40
2B Upper Floors	1,955	3.63	53 m²	36.85	0.46
2C Roof	1,665	3.10	7 m²	238.26	0.39
2D Stairs	90,610	168.42	510 m²	177.67	21.29
2E External Walls	85,105	158.19	561 m²	151.70	20.00
2F Windows and External Doors	6,780	12.60	12 m²	564.90	1.59
2G Internal Walls and Partitions	9,135	17.51	23 m²	110.28	2.22
2H Internal Doors	11,010	20.47			2.59
3. Finishes					
3A Wall Finishes	3,145	5.85			0.74
3B Floor Finishes	4,405	5.29	56 m²	50.82	1.04
3C Ceiling Finishes	7,865	14.62	56 m²	140.46	1.85
4. Fittings and Furnishings	4,090	7.60			1.12
5. Services					
5A Sanitary Appliances	9,865	18.34	5 No	1,937.40	2.32
5G Ventilating System	3,145	5.85			0.74
5H Electrical Installation					
5I Gas Installation					
5J Lifts and Conveyors	6,290	11.69			1.48
5K Protective Installation	37,750	70.17			8.87
6A Site Works	42,055	78.17			9.88
6. External Work Total	42,055	78.17			9.88
Preliminaries	22,020	40.93			10.58
Project Cost	£425,560	745.38			100.00

Review Question

The obvious areas for investigation are those high-cost elements except that when we review the elemental list provided earlier in Table 8.7 there are very few elements that represent a high proportion of the total cost. Two of the highest cost elements (2D Stairs and 2E External Walls) are highlighted in Table 8.12.

The remaining elements do not appear excessive in percentage of the total cost. In fact, the project appears to be quite basic with few obvious areas for cost reduction. However, one high-cost element does seem to stand out. The cost of 2D Stairs does appear to be quite high in terms of total cost (£90,610) and as a percentage of the total cost 21.29%. This element should be reviewed carefully as a priority for the next stage (Final Proposals) with a view to making a reduction matching at least the reduction we have to make to bring the cost within budget. If a greater saving can be made then this should be pursued and incorporated in to the next cost plan.

A review of cost analyses of similar projects to check on the percentage and proportion of the budget allocated to stairs will also give the design team the necessary background to make the changes in line with a balanced allocation of costs between elements.

Chapter 9

Cost Planning the Tender Documentation Stages

> The paramount management responsibility in the development phase
> is to ensure that design converges on the cost goals rather than allow cost
> to converge on design during development. This means that essential
> functionality becomes the design objective, and that design efficiency must
> be maximised ... Problem solutions that violate cost goals should not be
> tolerated unless essential functionality and quality are in jeopardy ...
>
> Michaels and Wood (1989: 23, 24)

Chapter Preview

Tender Documentation is the overall or inclusive title given to the final two design stages: Final Proposals (Work Stage E) and Production Information (Work Stage F). These are the stages where design decisions are concluded to form part of the project that will eventually go out to tender. If incorrect decisions are made that lead to the budget being exceeded then the documentation will have to amended, either before the contract is signed or as part of a variation, or variations, to reduce the contract sum to within the budgeted amount. Obviously, if the project documentation has to be amended this is not a satisfactory state of affairs and can be measured as a failure on the part of the design team. Therefore, everyone in the design team should redouble their efforts in the Final Proposals and Production Information stages to prevent this unsatisfactory conclusion. If it means changing a favourite detail or cherished material without compromising the quality of the project, then it must be done.

The previous stages have all been concerned with planning both the design and the cost of the design. The Documentation Stage is the crucial stage in controlling the cost of the detailed or documented design. At the end of the Production Information stage the design team should have completed the detail design of the buildings in accordance with the decisions agreed during the Detailed and Final Proposals stages as a result of the detailed investigations carried out in the cost checking process.

The RIBA (2000) make an important point about the preceding stage of Detailed Proposals and this stage of Final Proposals in the objective of maintaining design and cost control. It is suggested that, 'a significant contribution to making the process efficient and cost effective ... if client and designers agree to freeze:

- The developed project brief at the end of the Detailed Proposals (Stage D), and

- The design ... at the end of the Final Proposals (Stage E).'

This chapter builds upon and continues with the applied aspects of the earlier chapters. In Chapter 8 we completed the Detailed Proposals Stage cost plan and in this chapter we take our project to its conclusion in the Final Proposals and Production Information stages where we produce the *Tender Document Stage Cost Plan*.

Using the (RIBA, 1998) framework diagram we introduced and developed in the three earlier chapters, the Documentation Work Stages E and F are shown shaded in Table 9.1.

Table 9.1 RIBA Outline Plan of Work 1998

Pre-design (Briefing)			Design			
Inception or Feasibility			Pre-construction period			
Pre-stage A	Work Stage A	Work Stage B	Work Stage C	Work Stage D	Work Stage E	Work Stage F
	1	2	3	4	5	6
Establishing the Need	Options Appraisal	Strategic Briefing	Outline Proposals	Detailed Proposals	Final Proposals	Production Information
Chapter 6			Chapter 7	Chapter 8	Chapter 9	
Briefing			Outline Proposals	Detailed Proposals	Tender Documentation	

Overview of Tender Document Cost Plan Activities and Tender Documentation

Whilst the RIBA (2000) Plan of Work expects this final part of the design work to be carried out in two separate stages the cost planning activities are integrated to produce a Tender Document Stage Cost Plan.

The project brief is agreed at the end of the Detailed Proposals and should not be modified at all during this stage unless the client requests and ratifies a change for valid reasons. Cost monitoring is carried out from the developing Production Information (mainly working drawings and specification). The Tender Document Cost Plan is prepared from the final working drawings, specification and other tender documents.

The cost plan does not focus entirely on the building costs alone. The external works must also be cost planned and checked with remedial action necessary as required. Similarly, engineering costs must also be integrated into the cost plan and other Production Information. As the work proceeds in this stage cost checks are incorporated for the engineering services and structural systems using the working, concept or shop drawings as they are developed by the respective disciplines.

The cost planner must effectively monitor cost by providing cost checks during Final Proposals and Production Information stages, which is mainly concentrated on the detailed design development and production of tender documents. This is a continuous process that includes comparative cost studies, research and advice on the economic effect of proposed modifications including the use of alternative finishes and methods of construction. Cost statements should be prepared and submitted to the design team on at least a monthly basis. In addition, cash flow forecasts and sustainability studies of any alternative passive design features and technological intervention to provide energy savings are likely to be required by the design team. This is the last opportunity to model and cost the effects of any of these changes on the elements being considered.

The final Tender Documentation Cost Plan must be comprehensive in its scope and presented on at least an elemental and in some cases, a sub-elemental basis incorporating all changes made as a result of cost checks during detail development and preparation of tender documents. It is important that sub-elemental unit rates and quantities to be shown to enable the design team to judge and assess the quantity of element they have incorporated into the project.

Work Stage E Final Proposals: Key Activities

The key design team activities at the Final Proposals stage that affect the Cost Planning function are:

- Completion of the final layout drawings and approved by the client.

- Agreed positions for service terminals, ceiling layouts and major builder's work in connection with building services installations.

- Complete sizing of all structural elements.

- Preparation and submission of all information required by statutory and other approvals for the project.

- Review of procurement method.

In terms of the cost planning activities the cost planner will be involved in include:

- Continuously check the cost of the design against the cost plan and advise the design team of the results and consequences (cost checking).

- Review design coordination with design or project manager or architect.

- Prepare updated cost plan as design information becomes available.

- Review and advice on procurement method.

- Prepare updated cash flow projections to design team and client.

- Revise cost plan as necessary, including any changes to the estimated total cost and amend cash flow as more detailed information is presented.

Finally, the Final Proposals design (Stage E) at the end of this first stage of the Tender Documentation is now frozen and no further changes can be made by anyone in the design team. The only permissible changes from now on will be client-initiated changes only.

Work Stage F Production Information: Key Activities

The Production Information continues the work established and decided in the Final Proposals and its main aim is to complete all the documentation of the agreed (or frozen) design. Specifically, the design team have to complete the following documentation:

- The architect (or designer) and the structural engineer have to coordinate, prepare and complete all the Tender Documentation consisting of location, assembly and component drawings, schedules and specification.

- The services engineer has to coordinate all the Production Information for the services systems consisting of schematic, detailed design and coordination drawings, builder's work and fixings schedules, specification design criteria and calculations.

The cost planning activities at this final stage continue the work of the previous stage, except greater emphasis is placed on taking remedial action to keep within the budget and to maintain the requirements of the finalised project brief and the agreed design.

Stages E and F Result: Tender Document Cost Plan

The cost planning activities are centred on the development and production of the Final Proposals for the project that eventually become the working drawings and associated documents (schedules, specifications and other project information) that eventually become the Production Information of the final stage. For our purposes and working as a cost manager on the cost plan we shall not distinguish between these stages because our cost planning activities are focused on the development and production of the Final Proposals for the project. In practice, most of this activity will primarily take place in the Final Proposals stage.

The purpose of these stages are to ensure that the overall detailed design is contained within the final budget and to ensure that the final projected cost does not exceed the cost limit previously defined at the Strategic Briefing stage B. Appropriate action must be taken if cost checks show that budget is being exceeded. To this end the Project Team must meet regularly and report on at least a monthly basis.

In some cases it may be useful that the Cost Plan is accompanied by a reconciliation statement comparing individual elemental totals, areas, etc. with those in the Detailed Proposals Cost Plan (DPCP) or with any interim Tender Document Cost Plan that may have been prepared. All differences, physical, qualitative and financial must be reconciled.

Documentation Produced

RIBA (2000: 4) states that this Design Stage E: Final Proposals is for the 'preparation of Final Proposals for the project sufficient for coordination of all components and elements of the project'. The final design stage F: Production Information is further sub-divided into two parts (RIBA, 2000: 4):

1. Preparation of Production Information in sufficient detail to enable a tender or tenders to be obtained and for applications for statutory approvals.
2. Preparation of further Production Information required under the building contract.

The Production Information (RIBA, 2000: 71) commonly produces the following documents on a fully designed building project:

- *Location drawings*: Drawings up to a 1:50 scale showing layouts and relationships. A typical location drawing is shown in Figure 9.1.

- *Assembly drawings*: Detail drawings, usually at 1:20 or 1:10 or larger scales, showing how different elements relate to each other. A typical assembly drawing is shown in Figure 9.2.

- *Component drawings*: Drawings giving information necessary for the manufacture of particular components.

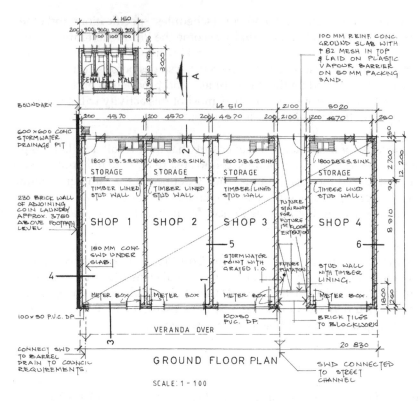

Figure 9.1 Location Drawing

- *Specifications*: Document with technical requirements of materials and standard of workmanship required on the project.

- *Bills of quantities*: Detailed schedule of all quantities of materials, labour and any other items required to execute a project. Its content is based on the drawings, specification and schedules prepared by the design team at the Production Information stage.

Typical location and assembly drawings from the Production Information stage are illustrated in Figures 9.1 and 9.2.

The Principles of Cost Control Incorporated During Tender Documentation Cost

The Final Proposals and Production Information stages must adopt the last two principles of cost control and management to guide the cost planning activities:

- a method of checking,

- a means of remedial action.

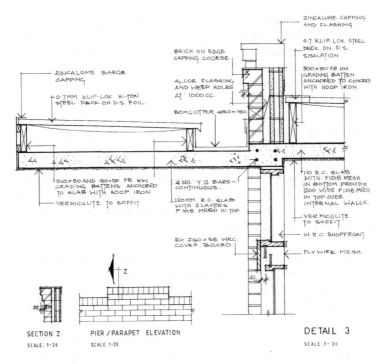

Figure 9.2 Assembly Drawing

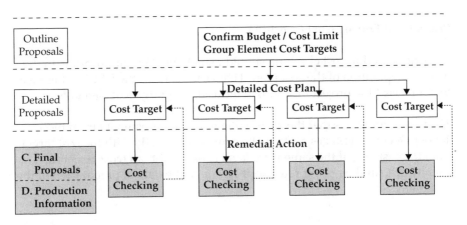

Figure 9.3 Cost Planning Concept Map: Detailed Proposals Stage
Source: Adapted from Smith and Love (2000: 11).

These are illustrated in our cost planning concept map in Figure 9.3.

Cost Checking

During these two stages the design team will be working on the final location and assembly drawings (working drawings) and details that will become part of the contract documents. The priorities for preparing and completing details should ideally follow the pattern that gives priority to the high cost and elements with a great deal of influence over other elements. These should be completed as early as possible in the Production Information stage. However, demands on resources and time may prevent the ideal pattern of operations during this stage.

The detailed design of each element should be cost checked as soon as it is completed and if necessary, remedial action should be taken.

Cost checking an element consists of estimating the probable cost of the element's detail design (which may be rough dimensioned sketches and specification notes) and comparing this with the cost target for the element from the Detailed Proposals Stage.

It is generally accepted that the most accurate method of estimating at this stage is the *Approximate Quantities* approach based upon a detailed or draft specification. Figure 9.4 shows the final stages of the cost planning process where the budget distributing techniques based on the elemental and approximate quantities methods dominate. Now that a full specification is, and should be, available for each element, this accepted traditional techniques of approximate quantities can be used with confidence. See chapter seven for an example of approximate quantities and references for further reading.

Prices Used During Cost Checking

The prices used in cost checking should usually be those appropriate to the date of preparation of the cost plan. That is, the date of the DPCP, where we first analysed the project in detail, in terms of cost targets of elements.

The reason for this is that the approximate quantities estimate from these Tender Document Stages can only be compared directly with the cost target from the cost plan if the same price level is used for both (cost targets are usually prepared by using rates appropriate to the date of preparation of the cost plan).

Remedial Action

There are two possible courses of remedial action which can be taken when cost checking shows that a cost target has been exceeded:

- the design of the element can be altered in an attempt to bring its cost within the cost target, or, as a last resort,

- the cost target for the element can be raised.

Pre-design			Design			
Inception or Feasibility			Pre-construction period			
Pre-stage A	Work Stage A	Work Stage B	Work Stage C	Work Stage D	Work Stage E	Work Stage F
	1	2	3	4	5	6
Establishing the Need	Options Appraisal	Strategic Briefing	Outline Proposals	Detailed Proposals	Final Proposals	Production Information
EARLY STAGE TECHNIQUES						
• Strategic value management		Cost limit	Cost targets		Cost checking	
• Strategic needs analysis						
• Quality function deployment						
• Situation structuring						
	• Expert choice					
		Traditional cost planning techniques				
	• Total cost					
	• Functional unit (student, bed, seat, room)		Budget distribution techniques			
		• Superficial (gross floor area)				
		• Functional area/cost				
			• Elemental (group)			
				• Elemental (Individual and sub-elemental)		
			• Approximate quantities (to suit level of detail available)			

Figure 9.4 Cost Planning Techniques and the RIBA Plan of Work

Naturally, we could have the situation where the cost target is higher than the cost checked element. In this pleasant case we may be able to act as benefactor to an element in distress, especially where the proposed element design is a sound and functional one, but insufficient funds are available from the initial cost target. In this case, the excess amount can be channelled to 'deserving' elements such as this one.

The project manager, design manager or architect has the responsibility of ensuring that action follows each cost check, whatever its results. At this late stage in the process it is essential that everyone is made fully aware of the decisions taken by formally documenting the changes. In Australia, the Australian Institute of Quantity Surveyors AIQS (2000) has recommended *Reconciliation Statement* (Table 9.2) and a *Cost Statement Summary,*

Table 9.2 Cost Plan Elemental Reconciliation Statements

AIQS Cost Planning Form 6 Reconciliation Statement – Element Summary					Job Ref. 08/99 March 2006
Code/Element	DPCP £	TDCP £	Increase £	Decrease £	Reasons for Movement
1 Substructure	40,900	46,310	5,410	–	Cost checked
2A Frame	14,470	–	–	14,470	Portals in External Walls element
2B Upper Floors	1,955				* to be cost checked
2C Roof	1,665				*
2D Stairs	90,610	58,000	–	22,610	Cost checked
2E External Walls	85,105				*
2F Windows and Ext Doors	6,780				*
2G Internal Walls and Ptns	9,435	6,000	–	3,435	Cost checked
2H Internal Doors	11,010	4,520	–	6,400	Cost checked
Etc					

*: to be cost checked. DPCP: Detailed Proposals Cost Plan; TDCP: Tender Document Cost Plan.
Source: Adapted from AIQS (2000: A1-14).

Reconciliation Statement Summary (Table 9.3) and are excellent formal methods of communicating this information to everyone in the design team and to the client.

There are three possible results and courses of action, which may follow the cost checking of an element summarised in Table 9.4.

If care has been taken to use the Detailed Proposals design and its cost plan targets as a basis for the design work during this Tender Document Stage, then action '3' above, *Confirm in writing that the design is suitable for working drawings*, should follow the cost checking of most of the elements.

The usefulness of the cost plan depends for a large part on the cost targets being realistic. Cost targets should, therefore, be altered only as a last resort, after every alternative design solution for a troublesome element has been rejected as unsuitable. Nevertheless, cost targets *may* have to be altered as users' requirements become better appreciated or have changed as a result of new information becoming available.

Table 9.3 Cost Plan Summary Reconciliation Statements

AIQS Cost Planning Form 4 Reconciliation Statement Summary – Single Building			Job Ref. 06/88 Date: March 2006
Project Title: *Offices*			
Building: *Single*			
Reconciliation between: DPCP and Tender Document Cost Plan (TDCP)			
Physical	**DPCP**	**TDCP**	**Reasons for Movement**
GFA	538 m^2	538 m^2	Increased/Reduced by ......*m^2 No change*
Area Efficiency	90%	90%	*Approximate*
Wall/Floor Ratio	1.47:1	1.47:1	
POP Ratio	73%	73%	
Other Physical Changes			*No change to shape or areas*
Costs			
Current Date	March 2006	March 2006	
Current BCIS All-In TPI	226	226	*Firm*
Tender Date	September 2006	September 2006	
Tender BCIS All-In TPI	231	231	*Projected until confirmed*
Building Cost		£338,485	
External Works		£42,055	
Preliminaries		£22,020	
Project Cost		£402,560	
Design and Contract Contingencies		£25,170	*Design (1.0%) and Contract Contingencies (2.5%) added*
Project Cost Tender		£427,730	
Other Costs			
Loose Furniture and Equipment		–	*No allowance for furniture and professional fees*
Professional Fees		–	
Contract Cost Adjustment		£8,560	*Allowance of 2.0%*
Gross Project Cost		£436,290	*Above budget of £425,000 (+£11,290)*

Source: Adapted from AIQS (2000: A1-16).

Table 9.4 Cost Checking and Courses of Action

Result	Possible Actions
• Cost of design *greater than* cost target	1. Amend design 2. Amend cost target
• Cost of design *equal to* cost target	3. Confirm in writing that the design is suitable for production drawings
• Cost of design *less than* cost target	4. Amend design (but unlikely) 5. Amend cost target(s)

Also refer to Figure 4.3 earlier for the pathways to final cost checking.

Cost Target Exceeded

If a cost target has to be raised there have to be compensating savings made in one or more elements to ensure the budget is not exceeded. To avoid exceeding the cost limit, cost targets should be adjusted throughout the cost plan to release surplus funds for the element in trouble.

This highlights another disadvantage of altering a cost target: even if the cost limit is not exceeded, the distribution of the available money over the various parts of the building will have to be changed. The result of this may be that the client cannot be given as good value for money as the design team planned during the preceding stages of the design process.

Cost Target Savings

If the estimated cost of the detailed design of an element turns out to be significantly less than the cost target, then we are faced with several possibilities. However, this is a pleasant problem to be faced with!

Possible actions are:

* *The cost targets for the other elements can be amended to absorb the surplus funds.*
  This would obviously be done if any of the previously designed elements were running into trouble at the cost checking stage. If the cost of the designs of the other elements were all nearly equal to their cost targets, however this action may disturb the balanced design allowed for in the cost plan.

  If there are many elements still to be designed and cost checked, on the other hand, it may be a good idea to hold on to the savings as an extra 'Design Risk' allowance in case some of the later elements run into trouble.

- *The probable savings can be notified to the client.*
  Let us be realistic! It would be a brave design team that chose to do this *before* receipt of the tender. If adequate standards have been provided throughout the building, the design team may well resolve to keep these savings as a bonus to be presented to the client on receipt of the tender.

- *The quality of the specification of the element can be increased to bring the cost up to the cost target.*
  Amending the design in this way is probably the most frequent action taken in this situation. The reason for this is that a cost plan attempts to achieve a balanced design; so increasing the quality of the specification of the element is more likely to give the client good value for money than by *returning* the savings to the client or by distributing them among the other elements. If the design team is happy with the quality of the detail design, it may well decide to *return* the savings to the client.

Question 9.1
Exceeding a Cost Target

Imagine our small offices project in the last chapter with its cost limit of £425,000. When three-quarters of the elements have been cost checked, the cost of the design is found to be running at £10,000 more than the sum of the cost targets concerned. The cost check of the next element (2C Roof) shows that the design is £7,000 over its cost target based on the specification given.

What should be done? The possible solutions may be:

(a) Amend its design to bring the cost down, or

(b) Increase the cost target and look for savings from the other elements.

BASIC CYCLE OF OPERATIONS

The basic cycle of operations during the Tender Document Stage is:

- Design
- Approximate quantities/measured estimate
- Comparison with cost target
- Decision and action.

This cycle of operations is repeated until all the elements have been designed and cost checked.

Question 9.2
Non-standard Elemental Qllowances

Which of the elemental allowances in a cost plan cannot be checked using the basic cycle of operations in this way?

Final Activity

Before leaving the Tender Document Stage Cost Plan, when all the elements have been designed and all the allowances checked once, a final cost check is performed to make sure that nothing has been overlooked.

Note that the drawings prepared during the Production Information stage for the Tender Document Stage, initially need not be polished working or Production Information drawings. They can simply be rough sketches provided that they contain all the information needed for the final and accurate production drawings, which should be prepared before the bills of quantities or other documents are measured. Ideally, working drawings should not be prepared for any element until after the final cost has been checked at the end of the Tender Document Stage. The reason for this is that it would be unwise to go to the expense of preparing working drawings for a design that may subsequently have to be changed, and the detailed design of an element may have to be altered at any time before the final cost check.

For instance, one of the last elements to be designed and cost checked may run into trouble, and consequently some of the earlier elements already designed may need to have the designs or cost targets altered.

The Cost Reconciliation Summaries/Statements shown earlier in Tables 9.2 and 9.3 are designed to keep an up-to-date record of the financial state of the project and to document changes and the decisions made.

A simplified account of cost checking has been given so far. Two refinements to this need to be considered and discussed:

- the order in which the elements should be designed and cost checked so as to avoid doing any unnecessary redesigning and rechecking,
- the best way to perform cost checking when the detail design is not produced one element at a time.

Question 9.3
The Order in Which Elements Should be Designed

Study the following table. Write down the major fault in the order of the cost checks. Briefly explain the possible repercussions of the fault.

Element	Cost Target £	Cost Check £	Date of Cost Check
2E External Walls	90,210	89,800	10 September 2007
2B Upper Floors	106,190	106,300	10 September 2007
2G Internal Walls	14,600	14,500	16 September 2007
2H Internal Doors	13,200	13,000	25 September 2007
2D Staircases	10,500	10,650	25 September 2007
2C Roof	49,710	49,800	25 September 2007
2F External Doors	77,270	77,000	4 October 2007
2A Frame	38,810	53,500	4 October 2007
3B Floor Finishes	69,100	69,000	12 October 2007
4 Fittings	4,020	4,500	12 October 2007

Another aspect of the unreliability of cost targets needs further consideration. For example, in the question above the cost targets given for 4 Fittings (£4,020) and 3B Floor Finishes (£69,100) are both unreliable because the design team has not considered the specification for these items as they can usually be safely left to later in the Production Information stage. The specifications of these elements do not usually affect the design of other elements.

In practice we must decide which of these elements Fittings or Floor Finishes should be designed and cost checked first. It is recommended that Floor Finishes should be designed and cost checked before Fittings. The reason for this is that changing a large cost target could give rise to embarrassingly large compensating changes to the smaller cost targets for other elements. Although cost targets should not be changed unless it is impossible to change the corresponding detail designs, this point is worth bearing in mind when planning the order in which the elements are to be designed and cost checked.

There are three characteristics of an element (a, b and c) that make it important to design the element early in the Documentation Stages, Final Proposals and Production Information:

(a)   Unreliable cost target

(b)   Large cost target

(c)   A design which affects the design of other elements.

The relative importance of these three characteristics varies from project to project, but the second one is less important than the other two, except when it is combined with one of the others. The elements that exhibit these characteristics will also vary from project to project. For example, in the offices project in the last chapter a review of the summary costs shown in Table 8.10, DPCP form, is likely to exhibit the characteristics a, b and c respectively in each of the elements as follows in Table 9.5.

Table 9.5 Element Characteristics

	Element	Characteristics				Element	Characteristics		
1	Sub-structure	a	b		5C	Disposal Installation	a		
2A	Frame	a	b	c	5D	Water Installation	a		c
2B	Upper Floors	a	b	c	5E	Heat Source	a		
2C	Roof		b	c	5F	Space Heating	a	b	c
2D	Stairs		b		5G	Ventilating System	a		
2E	External Walls		b	c	5H	Electrical Installation	a	b	c
2F	Windows and External Doors		b		5I	Gas Installation	a		
2G	Internal Walls and Ptns		b		5J	Lifts and Conveyors	a		
2H	Internal Doors		b		5K	Protective Installation	a		
3A	Wall Finishes		b		5L	Communications	a		
3B	Floor Finishes		b		5M	Special Installations	a		
3C	Ceiling Finishes		b	c	5N	BWIC Services	a		
4	Fittings and Furnishings	a			5O	Profit and Attendance	a		
5A	Sanitary Appliances	a			6	Externals	a	b	
5B	Services Equipment	a							

Question 9.4
Ordering the Priority Elements

Study the Table 9.5 and write down the name of the elements corresponding to:

   (a) the first four elements, which you would design;

   (b) the elements, which would have, second priority.

In practice there will always be several other considerations to be borne in mind when deciding the order in which the elements are to be designed and cost checked. The listing of priorities outlined above should be considered when a programme for the Final Proposals and Production Information stages (Documentation) is prepared at the start of the stages.

The order in which elements are designed and cost checked is important and target dates should be set for:

- the completion of each element's detail designs;
- the completion of each element's cost check;
- the completion of the action following each cost check.

If remedial action is necessary when any of the elements are cost checked, there is obviously going to be a lot of paperwork in circulation among the design team at any one time. To avoid confusion between original and amended versions of any item, all cost checks, decisions and alterations should be committed to the standard cost plan and reconciliation statements as recommended in Tables 9.2 and 9.3.

We will now consider the second point mentioned earlier: how to perform cost checking when the detail design is not produced element by element.

Cost Checking When the Design is Not Produced Element by Element

Suppose all the superstructure group of elements (1–6) were designed concurrently and completed at the same time. If all the elements were then cost checked, it is possible that they would *all* have to be redesigned because one or two of the important ones (such as the frame and upper floors) had overshot their cost target.

This approach can create problems with controlling costs and the best, but not ideal, way of dealing with this situation is to cost check each element several times during the design of the group by costing the designed part of the element and adding an allowance for the part not yet designed.

Although it may seem rather long winded, the best solution is to cost check each complicated element several times during its detail design.

Cost Checking of Significant Elements

Suppose elements 1 Substructure, 2A Frame, 2B Upper Floors, 2C Roof and 2E External Walls and External Doors were being designed concurrently. The first cost check could be performed when approximately 25–30% of each element has been designed.

It is relatively easy to prepare an Approximate Quantities estimate for the designed part of each element. The only slight difficulty is in deciding allowances for the parts of each element not yet designed. The estimate plus the allowance for each element is then compared with the corresponding cost target, and action taken. Although this process is more trouble (and probably less accurate) than designing and checking one element at a time, it nevertheless gives valuable information to the design team in time for the designs to be changed without undue trouble.

Example of Cost Checking

The following example is a cost check of the 2H. Internal Doors cost target for a hypothetical office project with Loading Bay. The DPCP allowances for 2H. Internal Doors is £44,030 and the sub-elemental allowance for external doors is £25,000.

The design team decide to use 40% hollow core doors (10 No. from a total of 25 No.) instead of wholly solid doors throughout (except for toilets). The remaining 21 No. internal doors have not been considered in detail yet and are left as the previous allowances.

Similarly, external doors have only been partially detailed, but early details indicate that an additional 1 No. solid core door from a total of 4 No. doors will have to be glazed for entry standard doors. This change means there are now 2 No. total glazed doors and 2 No. solid doors. The external fire door and roller shutter door have not been considered yet.

Note that changes in price level are not considered in this project in order to simplify the points made. The sources for the cost information in pricing each of the items listed could be one of the standard published price books such as *Spons*, *Laxtons* or *Rawlinsons*, materials and component suppliers, subcontractors or basic estimating calculations.

Office Project with Loading Bay

2H. Internal Doors   April 2006
   15 No. Doors changed from hollow core doors to solid
   core doors.
   Remaining doors as the previous Cost Plan Allowance.
   Cost Plan Allowance from DPCP = $44,030

Cost Check No. 1   Drawing A/86/25–6

Revised details   £

- Solid core timber doors   15 No. @ £303.00   4,545
- Hollow core doors   10 No. @ £325.00   3,250
                                                                      ———————
                                                                      £7,795

Remaining doors as Cost Plan Allowance `

• Laminate faced hollow core toilet doors	6 No. @ £238.00	1,428
• Fire-Rated Doors – 1 h	7 No. @ £1,511	10,577
– 2 h	5 No. @ £1,827	9,135
– 3 h	3 No. @ £2,412	7,236
• Panels	18m2 @ £233	4,194
Cost check No. 1 Total		£40,365
Cost Plan Allowance: 2H		£44,300
Net saving: 2H		−£ (3,935)

Action

The saving in this element $3,935 is then carried forward to the 2F.2 Sub-Element External Doors where changes will have a flow on cost effect.

Office Project with Loading Bay	April 2006
2F.2 Sub-Element: External Doors	Drawing A/86/25-7

Cost Check No. 2
1 No. Doors changed from solid core doors to glazed doors (=2 No. solid, 2 No. glazed).
Remaining metal clad and roller shutter doors as the previous Cost Plan Allowance.
Cost Plan Allowance from DPCP = $25,000

Revised details

• Solid core doors	2 No. @ £1,115	2,230
• Glazed entry doors	2 No. @ £2,800	5,600

Remaining doors as Cost Plan

• Metal clad fire doors	1 No. @ £3,600	3,600
• Roller shutter doors	1 No. @ £15,530	15,530
Cost check No. 2 Total		£26,960
Cost Plan Allowance: 2F.2		£25,000
Net increase: 2F.2		+£1,960

Action decided

Net effect: 2H	−£3,935
2F.2	+£1,960
Net saving from cost checks	−£1,975

This is good news and this saving will assist in providing additional funds in other elements, if required. Therefore, we can confirm drawings A/86/25-6 and 7 as suitable for Production Information. These elements should proceed into further design development.

It is essential to have a well-prepared programme of work for the Production Information Work Stage and members of the design team should keep in regular contact. Updates on the cost plan should be regularly circulated and updated as layout, assembly and services drawings, schedules and specifications are produced during this final stage of cost planning.

Summary of Cost Control During the Tender Document Stage

- Each element must be cost checked as soon as its design is completed, at the least.
- If several elements are designed concurrently, they should all be cost checked several times during their design.
- Whatever the result of a cost check, action *must* be taken.
- There should be a carefully prepared programme of work for this stage.
- There should be frequent communications between the members of the design team.
- Standard Cost Plan Reconciliation Forms should be used to communicate, date and record all cost decisions for the benefit of the client and the design team.

Chapter Review

The Final Proposals and Production Information stages (Tender Documentation) are the most important stages of the whole project. It is here that the final decisions will be made as to the nature and extent of the content of each element. The role of the cost planner is not one of merely pricing the decisions of the design team, but must be intimately involved in the design discussions and decision-making process. It requires active participation on the part of the cost planning team and demands an extensive knowledge of materials and characteristics for the various decisions being made. The decisions cannot be viewed in isolation and the interaction between the decisions made for specific elements must be recognised and communicated before a final decision is made. It requires a high level of activity, expertise, tact and ingenuity to arrive at a balanced approach to the final design. Unfortunately, this rich complexity of interaction cannot be portrayed accurately in this chapter.

Whilst the description given in this chapter may give the impression that cost checking is repetitive and mechanical, this is unlikely to be the case in practice. If the process is conducted as merely accounting for decisions, then the cost checking and design advice is not being carried out adequately or rigorously.

The decisions made during the Tender Document Cost Plan stage must also not have the single aim of minimising the capital costs of the project. To labour the point again, and to return to the theme of the earlier chapters, the time, cost and quality factors must be considered. In addition, decisions by the design team must include the broader issues of the value and life cycle cost effects of their decisions. A discussion of value in projects was introduced in Chapter 1. Refresh your memory of these concepts now, and in practice, as they are important. Neglect of these two important areas may disadvantage the overall value, the performance, income and running cost factors in the completed project for its whole life. Therefore, in some cases it may be more appropriate to allow a more costly option if it can be justified on performance and value grounds. However, the point must be stressed that good design and sound performance does not always have to be more expensive. The individual conditions and situation must be considered factually on their merits.

In this chapter we bring the cost planning activities to a close. We have shown you how to carry out cost checks in the in-text questions and you now have the knowledge to continue this process to its conclusion. Remember, there is no single answer on any project. You must take account of all the factors and circumstances that apply. At the end of the Tender Documentation stage you will have to prepare a final statement of cost to the client. This final report to the client is very important. You must ensure that the client's needs and requirements receive the highest priority and that the simple expediency of expecting the client to fund the budget excess should not be automatically assumed. This approach should be followed in this project and any others you are involved in at any time in the future.

This final stage calls upon your communication and cost planning skills within the design team to anticipate and advise on relevant solutions. The technical answer to the elemental choice must be analysed, discussed, argued if necessary, re-analysed, if necessary and then documented and communicated to all members of the team. It must be remembered that copious and complex documentation that has been described in this, and earlier chapters may not impress the client. As a cost planning professional you must be able to communicate the design team's recommendations, decisions and actions clearly to the client and to any other decision-maker. It is pointless to arrive at the most sensible balance of time, cost and quality in a project if it is still not acceptable to the client.

Reflections on the Accuracy of Forecasting Estimates

Research in the UK by Ashworth and Skitmore (1982), Beeston (1983), Skitmore (1988), the University of Reading (1980a, b, 1981) indicates that the accuracy of cost planning techniques naturally reflects three important factors:

- quality of project information;
- extent of project information;
- type and extent of the cost database.

The extensive study by the University of Reading (1980a, b, 1981) presented a summary of the accuracy of the commonly used techniques in modern cost planning. These results were gained by a computer based simulation model of the environment in which the cost planner works and the results are summarised in Table 9.6.

When elemental estimating is based on a sound statistical analysis of all relevant databases (item 5 in Table 9.6) improvements in accuracy can clearly be detected with a mean deviation of tenders of 6%. These results support the need for the Building Cost Information Service (BCIS) to continue with its collation of cost data and analysis of projects as it provides the cost planner with the ability to improve cost planning accuracy to predict tender amounts.

It is also interesting that whilst most quantity surveyors or cost planners would suggest that the final accuracy of their estimates should be within ±5% of the tender figure, the Table 9.6 suggests that this level of accuracy

Table 9.6 Quantity Surveyors' Estimating Accuracy in Theory

Estimating Method	Mean Deviation of Estimates from Tenders (%)	Coefficient of Variation of Errors (%)
1. Cost per square metre taken from one previous project.	18	22.5
2. Cost per square metre derived by averaging rates from a number of previous projects.	15.5	19
3. Elemental estimating based on rates taken from one project.	10	13
4. Elemental estimating based on rates derived by averaging the rates taken from a number of previous projects.	9	11
5. Elemental estimating based on statistical analysis of all relevant data in the database.	6	7.5
6. Resource use and costs based on contractors' estimating methods.	5.5	6.5

Source: Property Services Agency (UK) (1981: 25).

would be difficult to achieve consistently even when using the most detailed of the estimating methods; contractor's estimating methods. In a practical study of the estimating accuracy of quantity surveyors when compared with the lowest tender, the average accuracy was only 9.8%. The report is a sobering evaluation of the cost planning and estimating techniques that cost planners use.

Two major factors are noted in the report as contributing to the discrepancy of approximately 12% between the tenders and the estimates and the target of 5% based on the variability in the lowest tenders. These factors are:

- The addition of large lump sums in building up the total estimates for preliminaries, inflation, locality allowance, specification level and professional intuition.
- The choice of items or elements to measure and price is important because a large percentage of the total cost is attributable to a relatively small number of items or elements. That is, concentrate your efforts on the few items that have real cost significance and the accuracy of the estimate is thereby improved.

For this reason when choosing a percentage for the lump sum additions for design risk and contract contingencies mentioned in this and earlier chapters if poor judgement is applied it can negate all the good work that may have been put into estimating the rest of the work. The second factor has been emphasised particularly in the last two chapters.

There is a danger from our descriptions of cost planning activities in Chapters 5–10 that we may have tended to see the process as linear, sequential and divided into the six (plus one) RIBA (2000) neat steps of pre-design Work Stages A and B, and the design stages C, D, E and F. These stages portray the cost planning process as a model of information development that represents a series of decisions as a single reporting point. The process is more than a series of results contained in a report. However, it is important to record decisions made and the cost consequences of those decisions for the client and the design team in a formal document.

Many distinct and cumulative parts exist in the life of a project that should not interrupt the continuity of the activities and the process. It is a continuous, interacting system that does not necessarily reach the end of the Documentation Stage in a strictly linear fashion. As decisions are made and inevitably changed and the design is adapted and re-formed the process takes on the characteristics of a highly active iterative system. After working through a complete stage that results in establishing a budget a client cannot afford, or a cost overrun on the established budget during a later activity, then the decisions will have to be revisited, possibly several times during the process. Cost planning is a demanding, interactive process that must involve all members of the design team and the client so that decisions can be evaluated and integrated properly. To labour the point, yet again, each decision will have cost and time consequences, in addition to the immediate quality considerations that all design decisions embody. The cost planner should be at the forefront of making a comprehensive assessment of the effect(s) of any design change on the project.

The techniques that we have described in Chapters 6–10 and summarised in Figure 9.4 (which was first presented earlier as Figure 8.1) tend to be viewed in isolation or in a compartmented way. In addition, the user of the techniques in practice may consider them as a passive collection, collation and documentation of cost data for each stage. The cost planner who is integrated into the design team will use these techniques in a positive and active way to guide and influence decisions. As the design progresses, the potential to advise and guide the design team expands because a wider range of techniques is available. The techniques available later in the process incorporate a larger number of design and other variables in their cost forecasting. As more design information is processed by these techniques it is possible to predict the total costs more accurately.

It is essential that the cost planner should be viewed as a useful and significant member of the design team who can provide essential management and advice of the cost, time and quality effect of the design team's decisions. Equally, the cost planner must consider the process as an integrated whole where all decisions, no matter how large or small they may be, are considered as having the potential to influence many parts of the project. For cost planning to succeed, it must involve a great deal of focused, productive activity supported by all participants in the design process.

It must never be forgotten that the systems and processes are established to *serve the needs and requirements of the client* and not to satisfy the whims and fancies of any specific design team member.

Review Question
Improving the Accuracy of Early Stage Estimates

Study Table 9.6 and then answer the question below.

What problems exist with the information found during the early stages of the design process? Propose methods by which the forecasting accuracy of traditional techniques can be improved?

The conclusion of the pre-construction process: post tender stages

In Chapter 6 we described the cost planning process in the simple sequential stages of:

- Setting the budget,
- Preparing the outline cost plan,
- Preparing the detailed cost plan,
- Cost checking and remedial action.

These stages have been described in Chapters 6, 7, 8 and this chapter. We are now left with the final stages that complete the pre-design and design stages of the RIBA (2000) Plan of Work. These two stages are:

- Stage G: Tender Documentation, and
- Stage H: Tender action (analysis of tender submission).

The tender stage is an important destination on our journey to complete the project and reaching this point with our budget intact will depend on the success of each of the Work Stages leading to it. The process of cost planning summarised above captures our three principles of cost control:

1. There must be *a frame of reference* (cost limit or budget) at the outset.
2. There must be *a method of checking* (cost target).
3. There must be *a means of remedial action* (cost check).

If we have conscientiously followed these principles throughout the Work Stages then the possibility of gaining a tender within or close to our original budget will be increased, but not guaranteed. The tender stage provides instant feedback on measuring the success or failure of our process.

Stage G: Tender Documentation
The aim of this stage is to prepare and collate all the Tender Documentation prepared in the Production Information stage and this documentation should be in sufficient detail to enable accurate tenders to be submitted for our construction project.

The conclusion of the Production Information stage is the completion of all the tender documents required for the Tender Documentation Stage. These tender documents will be issued to all tenderers and will form the basis for the cost planner/quantity surveyor to price these documents and prepare a final pre-tender estimate. This estimate should confirm the original project budget, or any amended value agreed to by the client during the design and cost planning stages. If this estimate produces a forecast tender amount well above, or well below, the budget then the cost planner and the project manager or design leader should prepare a strategy to cope with this situation if the tender is confirmed as a similar value to this final pre-tender estimate. Naturally, if the budget, tender and the final estimate are similar no action is likely to be needed. However, if there are large differences then the contingency strategy will have to be decided and implemented quickly.

Stage H: Tender action

The aim of this stage is to identify and evaluate potential contractors (and any specialists, if necessary) for the successful construction of the project. An important activity is gaining and appraising the tender submissions and then making a recommendation to the client as to which tender to accept. That is, the analysis of the tender submissions is a crucial step in the whole cost planning process.

This stage specifically involves (RIBA, 2000: 26,27) the following steps:

- Dealing with any pre-tender enquiries from potential tenderers.
- Carrying out checks on potential contractors.
- Confirming tenderers to the client.
- Providing tender documents to all selected tenderers.
- Answering all tenderers' queries during tender period.
- Receiving and opening tenders.
- Appraising all valid tenders.
- Carrying out any negotiations with the lowest and any other tenderers with the approval of the client.
- Preparing and submitting to the client the Tender Report, recommending the tender action required.
- Preparing any pricing documents for the adjustment of the tender sum and agreeing the revised tender amount with the approved contractor (if required).

When a tender is accepted the successful tenderer is transformed into a contractor once the building contract is signed. The RIBA (2000) Plan of Work then officially moves into the construction stages of:

- Stage J: Mobilisation.
- Stage K: Construction to Practical Completion and the final stage.
- Stage L: After Practical Completion.

The whole process of procuring a building project is shown in Table 9.7. It is the aim of this book that when the various cost planning stages (described in all the earlier chapters) in the Pre-design, Feasibility and Pre-construction have been rigorously and professionally carried out it should lead to the successful conclusion of the project.

Table 9.7 RIBA (2000) Plan of Work

	Inception or Feasibility	
Pre-design (Briefing)	Pre-Stage A	• Establishing the Need
	Work Stage A	• Options Appraisal
	Work Stage B	• Strategic Briefing
	Design	
Pre-construction period	Work Stage C	• Outline Proposals
	Work Stage D	• Detailed Proposals
	Work Stage E	• Final Proposals
	Work Stage F	• Production Information
	Work Stage G	• Tender Documentation
	Work Stage H	• Tender Action
	Construction	
Construction period	Work Stage J	• Mobilisation
	Work Stage K	• Construction to Final Completion
	Work Stage L	• After Practical Completion

References

AIQS (2000) *Australian Cost Management Manual*, AIQS, Canberra.

Ashworth, A. and Skitmore, R. M. (1982) *Accuracy in Estimating*, Chartered Institute of Building, Ascot, UK.

Beeston, D. (1983) *Statistical Methods for Building Price Data*, E. & F. N. Spon, London, UK.

Laxtons Building Price Book (2006) Johnson, V. B. (editor), Elsevier Press, Oxford.

Michaels, J. V. and Wood, W. P. (1989) *Design to Cost*, John Wiley and Sons, New York.

Property Services Agency (1981) *Cost Planning and Computers*, Her Majesty's Stationery office, London, UK, pp. 17–31.

Rawlinsons (annual publication) *Rawlinsons Australian Construction Handbook*, The Rawlinsons Group, Sydney.

Royal Institution of British Architects (1998) *Architect's Handbook of Practice Management*, 6th edition, RIBA Publications, London, UK.

Royal Institution of British Architects (2000) *The Architect's Plan of Work*, RIBA Publications, London, UK.

Skitmore, R. M. (1988) *An Empirical Study of Human Performance in Early Stage Construction Contract Price Forecasting*, Department of Civil Engineering, University of Salford, UK.

Smith, J. and Love, P. (2000) *Building Cost Planning in Action*, University of New South Wales Press, Sydney, Australia.

Spon's Architect's and Builders' Price Book (2006) Davis Langdon (editor), E & F N Spon, London, UK.

University of Reading (1980a) *Construction Cost Data Base – First Annual Report*. A report produced on behalf of the Property Services Agency by the Department of Construction Management, University of Reading, PSA Library, London, UK.

University of Reading (1980b) *Construction Cost Data Base – Second Annual Report*. A report produced on behalf of the Property Services Agency by the Department of Construction Management, University of Reading, PSA Library, London, UK.

University of Reading (1981) *Construction Cost Data Base – Third Annual Report*. A report produced on behalf of the Property Services Agency by the Department of Construction Management, University of Reading, PSA Library, London, UK.

Further Reading

Ferry, D. J., Brandon, P. S. and Ferry, J. D. (1999) *Cost Planning of Buildings*, 7th edition, Blackwell Science, Oxford.
    Chapter 17: Cost Planning at Production Drawing Stage, pp. 281–292.

Flanagan, R. and Tate, B. (1997) *Cost Control in Building Design*, Blackwell Science, Oxford.
    Chapter 7: Cost Control During Detail Design pp. 73–84.
    Chapter 12: Cost Control During Detail Design Stage pp. 283–300.

Jaggar, D. Ross, A., Smith, J. and Love, P. (2002) *Building Design Cost Management*, Blackwell Publishing, Oxford.
    Chapter 11: Confirming the Cost Targets: Detailed Design Stage pp. 134–140.

SUGGESTED ANSWERS

Question 9.1: Exceeding a Cost Target

(a) is more likely to be more correct than (b).
However, if it is known that the cost of the design could not be decreased without producing an element of unacceptably low quality, then (b) is obviously the only action possible.

Question 9.2

Non-standard elemental allowances
The obvious ones are:

- preliminaries
- contingencies.

These allowances cannot be checked in the same way as the other elements, but they should nevertheless be carefully considered during the Tender Document Stage.

The first two allowances can obviously be reviewed as the detailed design progresses, and any necessary changes made. However, the preliminaries element is a critical one. The earlier adjustments for *quantity, quality* and *price* level can rarely be made in the preliminaries element (00PR) of a cost plan. The preliminaries element is an important section of a cost plan and can represent from 5% to 20% of the total cost of the project.

Preliminaries
The assessment of the preliminaries element is influenced most by the type of construction and size of the building(s). However, more than any other element, preliminaries on each project contain a unique set of requirements giving each preliminaries element a special nature. In addition, the cost planner must also consider the overall impact of the construction programme, total project mass, size, area, building height, shape and closeness of adjacent buildings and activities on the final assessment of the preliminaries percentage.

See Chapter 7 for a discussion of preliminaries and their pricing.

Contingencies
The overall contract contingencies allowance may have to be changed if greater risk and uncertainty for the price levels in the contract as a whole occurred during the Tender Document Stage. However, it is normal that the

design risk is at its lowest by this stage and the allowance for this becomes zero by the completion of the Tender Document Cost Plan. The allowance for contingencies or allowance for any design risk in the elements themselves can be changed to reflect uncertain or changing conditions. Through these changes extra money can be supplied to any of the elements that may run into trouble during the cost checking process.

Question 9.3: Order of Cost Checks

The major fault is that the frame element (2A) is cost checked too late in the design process. The reasons for this being a fault are as follows:

(a) Judging by the difference between the cost of the design and the cost target, it was probably known during the Detailed Proposals Stage that this cost target was unreliable. All unreliable cost targets should be checked at the earliest date.

(b) Taking the remedial action of redesigning the frame may make it necessary to redesign many of the other structural elements. This would mean that most of the work that had gone into the designing of these elements would have been wasted. All elements whose design significantly affects the design of other elements should be cost checked before its related elements are designed.

Question 9.4: Priority of Cost Checks

The criterion which we hope you used is that elements which exhibit all three characteristics should probably be cost checked before those exhibiting two, and so on.

Working on this principle the first four to be considered are:

2A   Frame
2B   Upper Floors
5F   Space Heating
5H   Electrical Installation.

The next six are:

1   Substructure
2C   Roof
2E   External Walls
3C   Ceiling Finishes
5D   Water Installation
6   External Works.

In practice, there will always be several other considerations to be borne in mind when deciding on the order in which the elements are to be designed and cost checked. The approach we have outlined above should be carefully considered when the programme for the Production Information stage is prepared at the end of the Final Proposals stage or at the beginning of this stage.

An interesting theoretical extension of this idea could result in a prescriptive order for considering the elements using the following characteristics. As a reminder, the three characteristics of an element that would make it important to design the element early in the Production Information stage are:

(a) unreliable cost target,

(b) large cost target,

(c) a design that affects the design of other elements.

If only one characteristic is present at a time, the decreasing order of importance is probably (c), then a, followed by (b). When two characteristics are present then the decreasing order of importance is (c) + (a), then (c) + (b), followed by (a) + (b).

Review Question: Improving the Accuracy of Early Stage Estimates

Early stage estimates by their very nature are based on meagre information. Little information will have been developed about the physical characteristics of the project and to complicate the pricing process it may be difficult to identify appropriate cost data to use on the proposed project. In the traditional single rate estimating techniques that predominate at this stage (functional unit, superficial, functional space) there are two variables; number or area and a suitable rate to use in the technique. Naturally the result is sensitive to the rates and values used. In the later stages when we use floor areas, element and approximate quantities, more effort is needed in measurement. Some observers argue that the simpler the method for measuring then the more difficult it is to calculate a reliable and accurate price. This is supported by the material contained in the work of the Property Services Agency (1981) and researchers identified earlier in this chapter.

The difficulties faced by the cost planner and the design team in early stage estimates can be summarised under three headings of measurement, pricing and other factors. Suggestions for improvement in the accuracy of these estimates are offered:

- Measurement

  The cost planner should obtain appropriate quantifiable data from the client's brief or statement of requirements. This may be scanty and consist of little more than the building type, size or floor area and number of storeys. Specification information is unlikely to be known and the design team and the cost planner are likely to have to rely upon information from previous similar projects and from personal experience and knowledge of similar building types. Even when drawings are available these are likely to be in sketch or outline form and not totally reliable.

- Pricing

  The design team acting upon the advice of the cost planner must advise the client of the price at which a contractor is likely to be prepared to carry out the work. It is a prediction of the price of building work to be carried out at some time in the future. The cost planner will make a professional prediction of the price by reviewing recorded building cost data for similar work. It will be necessary to update and revise this to account for differences in time, location, market conditions and the design variables of height, shape, number of storeys and the factors discussed in Chapter 3.

  The cost planner is also attempting to predict the tender price for the project that itself is subject to a degree of error. Whilst some of the contractors' tender prices will reflect the true costs of the project, it is sometimes the case that some successful tenders are wildly inaccurate. Sadly, the contractor who makes the largest negative error on the tender is likely to win the contract. Therefore, the cost planner has a difficult task to predict the tender, particularly when this situation arises.

- Other factors

  Improve the briefing process by providing better quality and type of preliminary information supplied to the design team.

  Make better efforts to quantify the data that is available at the design stage. This goes beyond the element unit quantities and rates collected for cost analyses. It should also include design features such as number of storeys and shape indices. These factors should also be closely correlated with cost. The development of cost models and computer cost simulation models hold great promise for improving our ability to predict costs based on the barest of information.

  Extend and enhance the quality and type of cost information available to the cost planner. Whilst priced bills of quantities, cost analyses, price books of various kinds and estimating data are currently available more effort needs to be made in gaining access to a wider range of cost information through computer networks by collaboration in the industry. The BCIS is a long standing example of industry cooperation and building a community of interests amongst cost planners, clients, design team members and facilities managers.

Be more aware of estimating techniques used by the contractor's estimator and of the planning and construction techniques used by contractors. To advise on the time consequences of design decisions cost planners need to be familiar with programming and network techniques that model the construction of the project.

Appendix

STANDARD FORM OF COST ANALYSIS

Principles, Instructions and definitions

Standard Form
of
Cost Analysis

Principles, Instructions and
Definitions

 BCIS is a trading division of RICS Business Services Ltd

Standard Form
of
Cost Analysis

Principles, Instructions and Definitions

BCIS
3 Cadogan Gate
London SW1X 0AS

December 1969
(Reprinted February 2003)

CONTENTS

INTRODUCTION

The purpose of cost analysis is to provide data which allows comparisons to be made between the cost of achieving various building functions in on project with that of achieving equivalent functions in other projects. It is the analysis of the cost of a building in terms of its elements. An element for cost analysis purposes is defined as a component that fulfils a specific function or functions irrespective of its design, specification or construction. The list of elements, however, is a compromise between this definition and what is considered practical.

The cost analysis allows for varying degrees of detail related to the design process; broad costs are needed during the initial period and progressively more detail is required as the design is developed. The elemental costs are related to square metre of gross internal floor area and also to a parameter more closely identifiable with the elements function, i.e. the element's unit quantity. More detailed analysis relates costs to form of construction within the element shown by 'All-in' unit rates.

Supporting information on contract, design/shape and market factors are defined so that the costs analysed can be fully understood.

The aim has been to produce standardisation of cost analyses and a single format for presentation.

This document has been prepared jointly by J D M Robertson FRICS, AMBIM, on behalf of the RICS Building Cost Information Service and by R S Mitchell MRICS on behalf of the Ministry of Public Building and Works. The principles and definitions are based upon the report of a working party under the chairmanship of E H Wilson FRICS and the analysis of services elements has had the assistance of a report by a working party under the chairmanship of A W Ovenden FRICS.

The principles and definitions of cost analysis and this format are supported by:

> The Quantity Surveyor's Committee of the Royal Institution of Chartered Surveyors

> The RICS Building Cost Information Service, and

> The Chief Quantity Surveyors of:

>> The Ministry of Public Building and Works
>> The Department of Health and Social Security
>> The Ministry of Housing and Local Government
>> The Department of Education and Science
>> The Home Office
>> The Scottish Development Department.

THE STANDARD FORM OF COST ANALYSIS

1:   PRINCIPLES OF ANALYSIS

The basic principles for the analysis of the cost of building work are as follows:

1.1   A building within a project shall be analysed separately.

1.2   Information shall be provided to facilitate the preparation of estimates based on abbreviated measurements.

1.3   Analysis shall be in stages with each stage giving progressively more detail; the detailed costs at each stage should equal the costs of the relevant group in the preceding stage. At any stage of analysis any significant cost items that are important to a proper and more useful understanding of the analysis shall be identified.

1.4   Preliminaries shall be dealt with as prescribed for the appropriate analyses.

1.5   Lump sum adjustments shall be spread pro-rata amongst all elements of the building(s) and external works based on all work excluding Prime Cost and Provisional Sums contained within the elements.

1.6   Professional fees shall not form part of the cost analysis.

1.7   Contingency sums to cover unforeseen expenditure shall not be included in the analysis of prices, but shown separately.

1.8   The principal cost unit for all elements of the building(s) shall be expressed in £ to two decimal places per square metre of gross internal floor area.

1.9   A functional unit cost shall be given.

1.10   In Amplified Analyses, design criteria shall be given against each element. Special design and performance problems shall be identified.

1.11   The definitions of terms for cost analysis shall be those given hereafter.

1.12   The elements for cost analysis shall be those given hereafter.

1.13   The contents of each element shall be as given hereafter.

1.14   The principles of further detailed analysis shall be as given hereafter.

2: INSTRUCTIONS

2.1   GENERALLY

2.1.1   Definition of terms
Definitions of terms used throughout the analysis follow these instructions.

2.1.2   Complex contracts
A cost analysis must apply to a single building. In a complex contract (i.e. a contract which contains a requirement for the erection of more than one building) the size of the contract may have an important bearing on price levels obtained. If this situation occurs, it should be identified in the box 'Project details and site conditions' on the first sheet of the analysis.

2.1.3   Omissions or exclusions
Where items of work which are normally provided under the building contract have been excluded or supplied separately, this should be stated where appropriate.

2.2   PROJECT INFORMATION

2.2.1   Building type
CI/SfB classification will be given and restricted to the 'Built environment' classification taken from Table 0.

A 'College of further education' will, therefore, be classified and shown as:

CI/SfB			
722			
College of further education			

2.2.2   BCIS code
The BCIS reference code classifies buildings by the form of construction, number of storeys and gross internal floor area in square metres.

The different construction classes are:

A   Steel framed construction
B   Reinforced concrete framed construction
C   Brick construction
D   Light framed steel or reinforced concrete construction

A single-storey building of 766 square metres of gross internal floor area and built in traditional construction would have the following BCIS code:

C – 1 – 766

2.2.3   Client

Indication should be given of the type of client, e.g. borough council, church authority, owner–occupier, government department, property company, etc.

2.2.4   Location

The location of the project should be given, noting the city or the county borough or alternatively the borough and the county, e.g. Bristol, or Richmond, Surrey. The location may be reported less precisely if the client so desires.

2.2.5   Tender date

(1) Date fixed for receipt of tenders.

(2) 'The date of tender' i.e. 10 days before date of receipt.

2.2.6   Brief description of total project

(a) Brief description of the building being analysed and of the total project of which it forms part.

(b) Any special or unusual features affecting the overall cost not otherwise shown or detailed in the analysis.

2.2.7   Site conditions

(a) Site conditions with regard to access, proximity of other buildings and construction difficulties related to topographical, geological or climatic conditions.

(b) Site conditions prior to building, e.g. woodland, existing building, etc.

2.2.8   Market conditions

Short report on tenders indicating the level of tendering, local conditions with regard to availability of labour and materials, keenness and competition.

2.2.9   Contract particulars of total project

(a) Type of contract, e.g. JCT (with or without Quantities), GC/Wks/1, etc.

(b) Bills of quantities, bills of approximate quantities, schedule of rates.

(c) Open or selected competition, negotiated, serial or continuation contract.

(d) Firm price or, if fluctuating, whether for labour or materials or both.

(e) Number of contractors to which tender documents sent.

(f) Number of tenders received.

(g) Contract periods: (i) stipulated by client;
(ii) quoted by builder.

2.2.10 Tender list
(a) List of tenders received in descending value order.

(b) Indicate whether tenders were from local builders (L) or builders acting on a national scale (N).

2.3 DESIGN/SHAPE INFORMATION OF SINGLE BUILDINGS

2.3.1 Accommodation, design features
(a) General description of accommodation.

(b) Where a building incorporates more than one function (e.g. a block of offices with shops or car park deck) the gross floor areas of each should be shown separately.

(c) Where drawings are not provided, a thumbnail sketch shall be given of the building showing overall dimensions and number of storeys in height for each part related to ground floor datum (i.e. + for ground floor and upper storeys, − for basement storeys).

(d) Any particular factors affecting design/cost relationship resulting from user requirements or dictates of the site

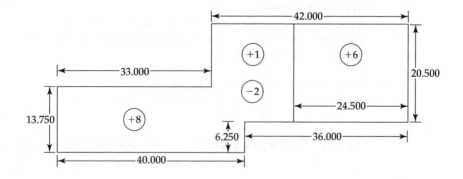

(user requirement is defined in RIBA Handbook as the area of accommodation, the activities for which a building is required, and the quality and standards it should achieve as stated by the client).

2.3.2   Floor areas
The measurement of floor areas and the information required on the form is that detailed hereafter under 'Definitions'.

2.3.3   Number of storeys
(a) Approximate percentage of building (based on gross floor area) having different number of storey, i.e. 20% single storey, 30% two storey, 50% three storey.

(b) Excludes structures such as lift, plant or tank rooms and the like above main roof slab.

2.3.4   Storey height
Storey heights shall be given and differing heights stated separately. (See Definitions.)

2.4   BRIEF COST INFORMATION

2.4.1   Preliminaries and contingencies
The totals for preliminaries and contingencies for the building being analysed should be stated and also each should be expressed as a percentage of the remainder of the contract sum. The analysis of prices does not include Contingencies.

2.4.2   Functional unit cost
The functional unit cost should be calculated by dividing the total of the group elements (including preliminaries but excluding external works) by the total number of functional units. (See Definitions.)

2.5   SUMMARY OF ELEMENT COSTS

2.5.1   Elements
The building prices are analysed by the elements.

Preliminaries are shown separately and also apportioned amongst the elements.

Where Preliminaries are shown separately each element is analysed under the following headings:

(a) Total cost of element.

(b) Cost per square metre of gross floor area £ and decimal parts of a £ (to two decimal places).

(c) Element unit quantity (as later described).

(d) Element unit rate £ and decimal parts of a £ (to two decimal places).

Where Preliminaries have been apportioned each element is analysed under the following headings:

(a) Total cost of element.

(b) Cost per square metre of gross floor area £ and decimal parts of a £ (to two decimal places).

2.5.2   Group elements

Sub-totals are shown for the group elements: substructure, superstructure, internal finishes, fittings and furnishings, services and external works.

Costs per square metre of gross floor area expressed in £ and decimal parts of a £ are calculated for the group elements.

Costs per square metre are also shown adjusted to a base date and in the case of analyses submitted to BCIS this will made by the Service using the BCIS cost indices.

2.6   AMPLIFIED COST ANALYSIS

2.6.1   Elements

The standard list of elements to be used for amplified cost analyses is that described in the following pages. The cost of each element must conform with the appropriate list of items shown in the specification notes for each element and with the principles of analysis.

2.6.2   Design criteria and specification

Design criteria relate to requirements, purpose and function of the element, and an outline of the design criteria is noted under each element.

The specification notes are considered to reflect architects' solutions to the conditions expressed by the design criteria and should indicate the quality of building achieved.

Specification notes provide a check list of the items which should be included with each element. Notes should adequately describe the form of construction and quality of material sufficiently to explain the costs in the analysis.

The instructions of the specification notes which follow are definitions of principle and where any departure from them seems necessary, a note should be made explaining how these cases have been dealt with.

2.6.3 Preliminaries

In the Amplified Analysis the element costs do not include Preliminaries which are to be analysed separately. However, under each element is to be included a figure which represents Preliminaries expressed as a percentage of the remainder of the contract sum.

2.6.4 External works

The expression of cost of external works related to gross floor area of the building(s) is not particularly meaningful. It is used in the 'Summary of element costs' so that the totals agree arithmetically, but in the Amplified Analysis there are no detailed costs required by this method.

2.6.5 Total cost of element

This is the cost of each element and the items comprising it should correspond with the notes in the right hand column headed 'Specification'.

If no cost is attributed to an element, a dash should be inserted in the cost column and a note made to the effect that this element is not applicable.

Where the costs of more than one element are grouped together, a note should be inserted against each of the affected elements, explaining where the costs have been included. For example, if windows in curtain walling are included in 'External walls' it should be so stated in the element 'Windows' and details of the cost included with the element 'External walls'.

2.6.6 Cost of element per square metre of gross floor area

This is the 'Total cost of element' divided by the gross floor area of the building.

2.6.7 Element unit quantity

In an amplified analysis, the cost of the element is expressed in suitable units which relate solely to the quantity of the element itself.

Instructions are given in the appropriate column which show what element unit quantity is to be used for each element, e.g. in the case of 'Floor finishes' the element unit quantity is the 'Total area of the floor finishes in square metres' and in the case of 'Heat source' the element unit quantity is 'Kilowatts'.

(i)   **Area for element unit quantities**
All areas must be the net area of the element, e.g. external walls should exclude window and door openings, etc.

(ii) **Cubes for element unit quantities**
Cubes for air conditioning, etc., shall be measured as the net floor area of that part treated, multiplied by the height from the floor finish to the underside of the ceiling finish (abbreviated to Tm^3).

2.6.8   Element unit rate

This is the total cost of the element divided by the element unit quantity. In effect it includes the main items and the labour items of the element expressed in terms of that element's own parameter. For example, in the case of 'Floor finishes' it is the total cost of the floor finishes divided by their net areas in square metres; in the case of 'Heat source' the elemental unit rate is the total cost of the heat source divided by its own parameter, the number of kilowatts. Elemental unit rates are shown in £ and decimal parts of a £ (to two decimal places).

2.6.9   Further quality breakdown

Where various forms of construction or finish exist within one element, the net areas and costs of the various types of construction should be included separately in the specification notes and provision has been made where appropriate.

The area of each form of construction is the net area involved and excludes all openings, etc.

The cost of each form of construction is the total cost of all items pertaining to that construction.

2.6.10   Example of amplified analysis layout

The following is an example of how the amplified analysis should be completed.

Element of design criteria	Total cost of element £	Cost of element per m² of gross floor area £	Element unit quantity	Element unit rate £	Specification
3.B **Floor finishes**	1,664	2.17	694 m²	2.40	19 mm granolithic laid monolithic, no skirting. 3 mm thermoplastic tiles Series 2 on 48 mm cement and sand screed, softwood skirting. 3 mm vinylised tiles on 48 mm cement and sand screed, softwood skirting. 25 mm (1″) 'West African' sapele wood block floor on 37 mm cement and sand screed, softwood skirting. 16 mm (5/8″) red quarries on 34 mm screed, quarry skirting.

Floor finishes	£	Area m²	All-in unit rate £
19 mm granolithic	30	30	1.00
3 mm thermoplastic Series 2	18	13	1.39
3 mm vinylised tiles	616	395	1.56
25 mm (1″) sapele blocks	570	161	3.54
16 mm (5/8″) quarries	430	95	4.52

Preliminaries 9.73% of remainder of contract sum.

3: DEFINITIONS

3.1 **Enclosed spaces**
1. All spaces which have a floor and ceiling and enclosing walls on all sides for the full or partial height.

2. Open balustrades, louvres, screens and the like shall be deemed to be enclosing walls.

3.2 **Basement floors**
All floors below the ground floor.

3.3 **Ground floor**
The floor which is the nearest the level of the outside ground.

3.4 **Upper floors**
All floors which do not fall into any of the previously defined categories.

3.5 **Gross floor area**
1. Total of all enclosed spaces fulfilling the functional requirements of the building measured to the internal structural face of the enclosing walls.

2. Includes area occupied by partitions, columns, chimney breasts, internal structural or party walls, stairwells, lift wells, and the like.

3. Includes lift, plant, tank rooms and the like above main roof slab.

4. Sloping surfaces such as staircases, galleries, tiered terraces and the like should be measured flat on plan.

   Note: (i) Excludes any spaces fulfilling the functional requirements of the building which are not enclosed spaces (e.g. open ground floors, open covered ways and the like). These should each be shown separately.

   (ii) Excludes private balconies and private verandahs which should be shown separately.

3.6 **Net floor area**
Net floor area shall be measured within the structural face of the enclosing walls as 'Usable', 'Circulation' and 'Ancillary' as defined below. Areas occupied by partitions, columns, chimney breasts, internal structural or party walls are excluded from these groups, and are shown separately under 'Internal divisions'.

1. **Usable**
   Total area of all enclosed spaces fulfilling the main functional requirements of the building (e.g. office space, shop space, public house drinking areas, etc.).

2. **Circulation**
   Total area of all enclosed spaces forming entrance halls, corridors, staircases, lift wells, connecting links and the like.

3. **Ancillary**
   Total area of all enclosed spaces for lavatories, cloakrooms, cleaners' rooms, lift, plant and tank rooms and the like, supplementary to the main function of the building.

4. **Internal divisions**
   The area occupied by partitions, columns, chimney breasts, internal structural or party walls.

   Note: The sum of the areas falling in the categories defined above will equal the gross floor area.

3.7   **Net habitable floor area (residential buildings only)**
   1.   Total area of all enclosed spaces forming the dwelling measured within the structural internal face of the enclosing walls.

   2.   Includes areas occupied by partitions, columns, chimney breasts and the like.

   3.   Excludes all balconies, public access spaces, communal laundries, drying rooms, lift, plant and tank rooms and the like.

3.8   **Roof area**
   1.   Plan area measured across the eaves overhang or to the inner face of parapet walls.

   2.   Includes area covered by rooflights.

   3.   Sloping and pitched roofs should be measured on plan area.

3.9   **External wall area**
   The wall area of all the enclosed spaces fulfilling the functional requirements of the building measured on the outer face of external walls and overall windows and doors, etc.

3.10   **Wall to floor ratio**
   Calculated by dividing the external wall area by the gross floor area to three decimal places.

3.11   **Element ratios**
Calculated by dividing the net area of the element by the gross floor area to three decimal places.

Note: In the case of buildings, where only a part is treated or served by mechanical or electrical installations, indication of this given by a ratio as follows:

$$\frac{t \text{ m}^2}{\text{gross floor area}}$$

where t m^2 is the total net area in square metres of the various compartments treated or served.

3.12   **Storey height**
1.   Height measured from floor finish to floor finish.

2.   For single-storey buildings and top floor of multi-storied buildings, the height shall be measured from floor finish to underside of ceiling finish.

3.13   **Internal cube**
1.   To include all enclosed spaces fulfilling the requirements of the building.

2.   The cube should be measured as the gross internal floor area of each floor multiplied by its storey height.

3.   Any spaces fulfilling a requirement of the building, which are enclosed spaces, such as open ground floors, open covered ways and the like, should be shown separately giving the notional cubic content of each, ascertained by notionally enclosing the open top or sides.

3.14   **Functional unit**
The functional unit shall be expressed as net usable floor area (offices, factories, public houses, etc.) or as a number of units of accommodation (seats in churches, school places, persons per dwelling, etc.).

4: FORMS OF ANALYSIS AND GUIDANCE NOTES

4.1   Concise, detailed and amplified forms of cost analysis

The standard method of cost analysis described here is in stages with each stage giving progressively more detail; the detailed costs at each stage should equal the costs of the relevant group in the preceding stage. At any stage of analysis any significant cost items that are important to a proper and more useful understanding of the analysis should be identified.

Forms of cost analysis have been prepared in three degrees of detail; Concise, Detailed and Amplified. The Detailed and Amplified forms are laid out as follows:

General and Background information and Summary of Elemental Costs for use with Detailed and Amplified Analyses.

Checklist of Specification and Design Notes which should accompany the Detailed Analysis.

Breakdown of the information required in an Amplified Analysis.

For details of the Concise Cost Analysis see the BCIS Section H.

4.2   Tank rooms

Where tank rooms, housing and the like are included in the gross floor area, their component parts shall be analysed in detail under the appropriate elements. Where this is not the case, their costs should be included as 'Builder's work in connection' (5.14).

4.3   Glazing and ironmongery

Glazing and ironmongery should be included in the elements containing the items to which they are fixed.

4.4   Decoration

Decoration, except to fair-faced work, should be included with the surface to which it is applied, allocated to the appropriate element, and the costs shown separately. Painting and decorating to fairfaced work is to be treated as a 'Finishing'.

4.5   Chimneys

Chimneys and flues which are an integral part of the structure shall be included with the appropriate structural elements.

4.6   Drawings

Drawings, A4 size negatives if these are available, should preferably accompany Detailed Cost Analyses.

TO BE USED WITH THE DETAILED AND AMPLIFIED ANALYSES

DETAILED COST ANALYSIS

CI/SfB

Job title:			
Location:	Client:		BCIS Code
	Tender dates:	(1)	(2)	Base month:

INFORMATION ON TOTAL PROJECT

Project and contract information

Project details

Site conditions:

Market conditions:

Project Tender
Price Index: Base:

Competitive tender list		
	£	N/L
Tender amended		

Contract particulars:

Type of contract: ....

Basis of tender*:   Open/Selected
competition

Bill of quantities   Negotiated

Bill of approximate   Serial
quantities

Schedule of rates   Continuation

Contract period stipulated by client........... months

Contract period offered by builder............. months

Contract period agreed................. months

Number of tenders issued.........

Number of tenders received...........

  * Tick as appropriate

Cost fluctuation*

		YES	?	NO
LABOUR		YES		NO
MATERIALS		YES		NO

**Adjustments based
on formula***   YES   NO

      * Tick as appropriate

Provisional sums   £...............

Prime Cost sums   £...............

Preliminaries   £...............

Contingencies   £...............

* Contract sum   £...............

* If this sum is not the same as the lowest tender
please state the reason why in the tender
amended box opposite.

TO BE USED WITH THE DETAILED AND AMPLIFIED ANALYSES   ANALYSIS OF SINGLE BUILDING

Design shape information

Accommodation and design features:

Areas

Basement floors	....	m²
Ground floor	....	m²
Upper floors	....	m²
Gross floor area	....	m²
Usable area	....	m²
Circulation area	....	m²
Ancillary area	....	m²
Internal division	....	m²
Gross floor area	....	m²
Floor spaces not enclosed	....	m²
Roof area...............		m²

Functional unit ....

$$\frac{\text{External wall area}}{\text{Gross floor area}} = \frac{}{} = $$

Internal cube = ...m³

Storey heights

Average below ground floor ....
at ground floor ....
above ground floor ....

Design/Shape

Percentage of gross floor area

a) Below ground floor ....%

b) Single-storey construction ....%

c) Two storey construction ....%

d) * storey construction ....%

e) * storey construction ....%

* Insert number of storeys.

Brief Cost Information

* Contract sum	£...............	
* Provisional sums	£...............	
* Prime cost sums	£...............	
* Preliminaries	£...............	being ....% of remainder of
* Contingencies	£...............	being ....% contract sum
* Contract sum less contingencies £...............		**(This sum should be the same as the total of the summary of element costs overleaf.)**
* Amounts for single building analysed.		

Functional unit cost .... Tender £...............

Excluding external works .... Base date £...............

TO BE USED WITH THE DETAILED AND AMPLIFIED ANALYSES

SUMMARY OF ELEMENT COSTS

Gross internal floor area: .... m²   Tender dates: (1)................ (2)................

Element	Preliminaries shown separately				Preliminaries apportioned amongst elements		
	Total cost of element £	Cost per m² gross floor area –£	Element unit quantity	Element unit rate £	Total cost of element £	Cost per m² gross floor area £	Cost per m² gross floor area (base date) £
1 Substructure	£	–£			–£	–£	£
2 **Superstructure**							
2.A Frame							
2.B Upper floors							
2.C Roof							
2.D Stairs							
2.E External wall							
2.F Windows and external doors							
2.G Internal walls and partitions							
2.H Internal doors							
Group element total	£	£			£	£	£
3 **Internal finishes**							
3.A Wall finishes							
3.B Floor finishes							
3.C Ceiling finishes							
Group element total	£	£			£	£	£
4 **Fittings and finishings**	£	–£			–£	–£	£

TO BE USED WITH THE DETAILED AND AMPLIFIED ANALYSES

5	**Services**					
5.A	Sanitary appliances					
5.B	Services equipment					
5.C	Disposal installations					
5.D	Water installations					
5.E	Heat source					
5.F	Space heating and air treatment					
5.G	Ventilating system					
5.H	Electrical installations					
5.I	Gas installations					
5.J	Lift and conveyor installations					
5.K	Protective installations					
5.L	Communication installations					
5.M	Special installations					
5.N	Builder's work in connection with services					
5.O	Builder's profit and attendance on services					
	Group element total	£	£		£	£
Subtotal excluding External works, Preliminaries and Contingencies		£	£		£	£
6	**External works**					
6.A	Site works					
6.B	Drainage					
6.C	External services					
6.D	Minor building works					
	Group element total	£	£		£	£
Preliminaries		£	–		–	£
TOTALS (less Contingencies)		£	£		£	£

TO BE USED WITH THE DETAILED ANALYSIS

SPECIFICATION AND DESIGN NOTES

Please include brief specification and design notes to describe adequately the form of construction and quality of material sufficiently to explain the prices in the analysis.

Check List

Group elements	Elements		Sub-elements	
1. SUBSTRUCTURE	1.A	Substructure		
2. SUPERSTRUCTURE	2.A	Frame		
	2.B	Upper floors		
	2.C	Roof	2.C.1	Roof structure
			2.C.2	Roof coverings
			2.C.3	Roof drainage
			2.C.4	Roof lights
	2.D	Stairs	2.D.1	Stair structure
			2.D.2	Stair finishes
			2.D.3	Stair balustrades and handrails
	2.E	External walls		
	2.F	Windows and external doors	2.F.1	Windows
			2.F.2	Externals doors
	2.G	Internal walls and partitions		
	2.H	Internal doors		
3. INTERNAL FINISHES	3.A	Wall finishes		
	3.B	Floor finishes		
	3.C	Ceiling finishes	3.C.1	Finishes to ceilings
			3.C.2	Suspended ceilings
4. FITTINGS AND FINISHINGS	4.A	Fittings and finishings	4.A.1	Fittings, fixtures and furniture
			4.A.2	Soft furnishings
			4.A.3	Works of art
			4.A.4	Equipment

TO BE USED WITH THE DETAILED ANALYSIS

5. **SERVICES**

5.A   Sanitary appliances
5.B   Services equipment
5.C   Disposal installations
    5.C.1   Internal drainage
    5.C.2   Refuse disposal
5.D   Water installations
    5.D.1   Mains supply
    5.D.2   Cold water service
    5.D.3   Hot water service
    5.D.4   Steam and condensate
5.E   Heat source
5.F   Space heating and air treatment
    5.F.1   Water and/or steam (heating only)
    5.F.2   Ducted warm air (heating only)
    5.F.3   Electricity (heating only)
    5.F.4   Local heating
    5.F.5   Other heating systems
    5.F.6   Heating with ventilation (air treated locally)
    5.F.7   Heating with ventilation (air heated centrally)
    5.F.8   Heating with cooling (air treated locally)
    5.F.9   Heating with cooling (air heated centrally)
5.G   Ventilating systems
5.H   Electrical installations
    5.H.1   Electric source and mains
    5.H.2   Electric power supplies
    5.H.3   Electric lighting
    5.H.4   Electric light fittings
5.I   Gas installation
5.J   Lift and conveyor installations
    5.J.1   Lifts and hoists
    5.J.2   Escalators
    5.J.3   Conveyors
5.K   Protective installations
    5.K.1   Sprinkler installations
    5.K.2   Fire-fighting installations
    5.K.3   Lightning protection

TO BE USED WITH THE DETAILED ANALYSIS

Group elements	Elements		Sub-elements	
	5.L	Communication installations		
	5.M	Special installations		
	5.N	Builder's work in connection with services		
	5.O	Builder's profit and attendance on services		
6. **EXTERNAL WORKS**	6.A	Site works	6.A.1	Site preparation
			6.A.2	Surface treatment
			6.A.3	Site enclosure and division
			6.A.4	Fittings and furniture
	6.B	Drainage		
	6.C	External services	6.C.1	Water mains
			6.C.2	Fire mains
			6.C.3	Heating mains
			6.C.4	Gas mains
			6.C.5	Electric mains
			6.C.6	Site lighting
			6.C.7	Other mains and services
			6.C.8	Builder's work in connection with external services
			6.C.9	Builder's profit and attendance on external services
	6.D	Minor building works	6.D.1	Ancillary buildings
			6.D.2	Alterations to existing buildings
PRELIMINARIES				
	Drawings: Drawings should accompany the Detailed Cost Analysis			

AMPLIFIED ANALYSIS

Element and design criteria	Total cost of element £	Cost of element per m² of gross floor area £	Element unit quantity	Element unit rate £	Specification
1.A **SUBSTRUCTURE** Permissible soil loading .... kN/m² (kilonewtons per square metre) Nature of soil .... Bearing strata depth .... m (metres) Site levels .... (to be given as gradient) Water table depth .... m (metres) Average pile loading .... kN (kilonewtons) Preliminaries .... % of remainder of contract sum.			Area of lowest floor measured as for gross internal floor area (m²).		All work below underside of screed or where no screed exists to underside of lowest floor finish including damp-proof membrane, together with relevant excavations and foundations. **Notes:** 1. Where lowest floor construction does not otherwise provide a platform, the flooring surface shall be included with this element (e.g. if joisted floor, floor boarding would be included here). 2. Stanchions and columns (with relevant casings) shall be included with 'Frame' (2.A). 3. Cost of piling and driving shall be shown separately stating system, number and average length of pile. 4. The cost of external enclosing walls to basements shall be included with 'External walls' (2.E) and stated separately for each form of construction.

Element and design criteria	Total cost of element £	Cost of element per m² of gross floor area £	Element unit quantity	Element unit rate £	Specification
2.A FRAME Grid pattern should be stated, giving centres of main columns in both directions.					Loadbearing framework of concrete, steel, or timber. Main floor and roof beams, ties and roof trusses of framed buildings. Casing to stanchions and beams for structural or protective purposes. **Notes:** 1. Structural walls which form an integral part of the loadbearing framework shall be included either with 'External walls' (2.E) or 'Internal walls and partitions' (2.G) as appropriate. 2. Beams which form an integral part of a floor or roof which cannot be segregated therefrom shall be included in the appropriate element. 3. In unframed buildings roof and floor beams shall be included with 'Upper floors' (2.B) or 'Roof structure' (2.C.1) as appropriate. 4. If the 'Stair structure' (2.D.1) has had to be included in this element it should be noted separately.
			Area of floors relating to frame, measured as for gross internal floor area (m²).		
Preliminaries .... % of remainder of contract sum.					
2.B UPPER FLOORS Design loads should be stated in kilonewtons per square metre (kN/m²) and spans given in metres.			Total are of upper floors (m²).		Upper floors, continuous access floors, balconies and structural screeds (access and private balconies each stated separately), suspended floors over or in basements stated separately. **Notes:** 1. Where floor construction does not otherwise provide a platform the flooring surface shall be included with

	£	Area m²	All-in unit rate £
Upper floors			

this element (e.g. if joisted floor, floor boarding would be included here).

2. Beams which form an integral part of a floor slab shall be included with this element.

3. If the 'Stair structure' (2.D.1) has had to be included in this element it should be noted separately.

Preliminaries .... % of remainder of contract sum.

2.C ROOF

Design loads should be stated in kilonewtons per square metre (kN/m^2) and spans given in metres (m).

The angle of pitch of sloping roofs shall be stated.

Area measured overall roof surfaces (m^2).

2.C.1 Roof structure

Construction, including eaves and verges, plates and ceiling joists, gable ends, internal walls and chimneys above plate level, parapet walls and balustrades.

Notes:

1. Beams which form an integral part of a roof shall be included with this element.

2. Roof housings (e.g. lift motor and plant rooms) shall be broken down into the appropriate constituent elements.

2.C.2 Roof coverings

Roof screeds and finishings. Battening, felt, slating, tiling and the like. Flashings and trims. Insulation. Eaves and verge treatment.

(contd.)...

Element and design criteria	Total cost of element £	Cost of element per m² of gross floor area £	Element unit quantity	Element unit rate £	Specification
2.C ROOF (contd.)...					**2.C.3 Roof drainage** Gutters where not integral with roof structure, rainwater heads and roof outlets. (Rainwater downpipes to be included in 'Internal drainage' (5.C.1).) **2.C.4 Roof lights** Roof lights, opening gear, frame, kerbs and glazing. Pavement lights.

Roof	£	Area m²	All-in unit rate £
Roof			
2.C.1 Roof structure			
2.C.2 Roof coverings			
2.C.3 Roof drainage			
2.C.4 Roof lights			

Preliminaries.............. %
of remainder of contract sum.

2.D STAIRS

The total vertical height of each staircase and its width between stringers should be given in metres (m).

2.D.1 Stair structure

Construction of ramps, stairs and landings other than at floor levels. Ladders. Escape staircases.

Notes:

1. The cost of external escape staircases shall be shown separately.
2. If the staircase structure has had to be included in the elements 'Frame' (2.A) or 'Upper floors' (2.B) this should be stated.

2.D.2 Stair finishes

Finishes to treads, risers, landings (other than at floor levels), ramp surfaces, strings and soffits.

2.D.3 Stair balustrades and handrails

Balustrades and handrails to stairs, landings and stairwells.

Stairs	£	Cost per m² of gross floor area £
2.D.1 Stair structure		
2.D.2 Stair finishes		
2.D.3 Stair balustrades and handrails		

Preliminaries ....% of remainder of contract sum.

Element and design criteria	Total cost of element £	Cost of element per m² of gross floor area £	Element unit quantity	Element unit rate £	Specification
2.E EXTERNAL WALLS $\dfrac{\text{External walls}}{\text{Gross floor area}} = \dfrac{\qquad \text{m}^2}{\qquad \text{m}^2} =$ $\dfrac{\text{Basement walls}}{\text{Gross floor area}} = \dfrac{\qquad \text{m}^2}{\qquad \text{m}^2} =$ The approximate value of thermal conductivity should be given in watt per metre degree Celsius (W/m°C). If calculated loadbearing brickwork, give indication of structural loading. Preliminaries ....% of remainder of contract sum.			Area of external walls measured on outer face (excluding openings) (m²).		External enclosing walls including that to basements but excluding items included with 'Roof structure' (2.3.1). Chimneys forming part of external walls up to plate level. Curtain walling, sheeting rails and cladding. Vertical tanking. Insulation. Applied external finishes. **Notes:** 1. The cost of structural walls which form an integral and important part of the loadbearing framework shall be shown separately. 2. Basement walls shall be shown separately and the quantity and cost given for each form of construction. 3. If walls are self-finished on internal face, this shall be stated.

		All-in unit rate £
	£	Area m²
External walls		

2.F WINDOWS AND EXTERNAL DOORS

$$\frac{\text{Windows}}{\text{Gross floor area}} = \frac{\underline{\hspace{2cm}} \, \text{m}^2}{\underline{\hspace{2cm}} \, \text{m}^2} = \underline{\hspace{2cm}}$$

$$\frac{\text{External doors}}{\text{Gross floor area}} = \frac{\underline{\hspace{2cm}} \, \text{m}^2}{\underline{\hspace{2cm}} \, \text{m}^2} = \underline{\hspace{2cm}}$$

$$\text{Area of opening lights to windows} = \underline{\hspace{2cm}} \, \text{m}^2$$

Preliminaries ....% of remainder of contract sum.

Total area of windows and external doors measured over frames (m²).

2.F.1 Windows
Sashes, frames, linings and trims.
Ironmongery and glazing.
Shop fronts.
Lintels, sills, cavity damp-proof courses and work to reveals of openings.

2.F.2 External doors
Doors, fanlights and sidelights.
Frames, linings and trims.
Ironmongery and glazing.
Lintels, thresholds, cavity damp-proof courses and work to reveals of openings.

Windows and external doors	£	Area m²	All-in unit rate £
2.F.1 Windows			
2.F.2 External doors			

Element and design criteria	Total cost of element £	Cost of element per m² of gross floor area £	Element unit quantity	Element unit rate £	Specification
2.G INTERNAL WALLS AND PARTITIONS $\dfrac{\text{Internal walls and partitions}}{\text{Gross floor area}} = \dfrac{\quad \text{m}^2}{\quad \text{m}^2} = \underline{\underline{\quad}}$			Total area of internal walls and partitions (excluding openings) (m²).		Internal walls, partitions and insulation. Chimneys forming part of internal walls up to plate level. Screens, borrowed lights and glazing. Moveable space-dividing partitions. Internal balustrades excluding items included with 'Stair balustrades and handrails' (2.D.3) **Notes:** 1. The cost of structural walls which form an integral and important part of the loadbearing framework shall be shown separately. 2. The cost of proprietary partitioning shall be shown separately stating if self-finished. Doors, etc., provided therein together with ironmongery, should be included stating the number of units installed. 3. The cost of proprietary WC cubicles shall be shown separately stating the number provided. 4. If design is cross-wall construction, the specification shall be stated and the cost shown separately.

		All-in unit rate £
Internal walls and partitions	£	Area m²

Preliminaries ....%
of remainder of contract sum.

2.H INTERNAL DOORS

Doors, fanlights and sidelights.
Sliding and folding doors.
Hatches.
Frames, linings and trims.
Ironmongery and glazing.
Lintels, thresholds, and work to reveals of openings.

Area of internal doors measured over frames (m²).

Internal doors	£	Area m²	All-in unit rate £

Preliminaries ....% of remainder of contract sum.

Element and design criteria	Total cost of element £	Cost of element per m² of gross floor area £	Element unit quantity	Element unit rate £	Specification
3.A **WALL FINISHES**			Total area of wall finishes (m²)		Preparatory work and finishes to surfaces of walls internally. Picture, dado and similar rails. **Notes:** 1. Surfaces which are self-finished (e.g. self-finished partitions, fair faced work) shall be included in the appropriate element. 2. Insulation which is wall-finishing shall be included here. 3. The cost of finishes applied to the inside face of external walls shall be shown separately.
					Wall finishes — £ — Area m² — All-in unit rate £
Preliminaries............% of remainder of contract sum.					
3.B **FLOOR FINISHES**			Total area of floor finishes (m²).		Preparatory work, screeds, skirtings and finishes to floor surfaces excluding items included with 'Stair finishes' (2.D.2), and structural screeds included with 'Upper floors' (2.B). **Note:** Where the floor construction does not otherwise provide a platform the flooring surface will be included either in 'Substructure' (1.A) or 'Upper floors' (2.B) as appropriate.
					Floor finishes — £ — Area m² — All-in unit rate £
Preliminaries............% of remainder of contract sum.					

3.C CEILING FINISHES

3.C.1 Finishes to ceilings
Preparatory work and finishes to surfaces of soffits excluding items included with 'Stair finishes' (2.D.2) but including sides and soffits of beams not forming part of a wall surface.
Cornices, coves.

3.C.2 Suspended ceilings
Construction and finishes of suspended ceilings.

Notes:
1. Where ceilings principally provide a source of heat, artificial lighting or ventilation, they shall be included with the appropriate 'Services' element and the cost shall be stated separately.
2. The cost of finishes or suspended ceilings to soffits immediately below roof shall be shown separately.

Ceiling finishes	£	Area m²	All-in unit rate £
3.C.1 Finishes to ceilings			
3.C.2 Suspended ceilings			

Total area of ceiling finishes (m²).

Preliminaries ....% of remainder of contract sum.

Element and design criteria	Total cost of element £	Cost of element per m² of gross floor area £	Element unit quantity	Element unit rate £	Specification
4.A FITTINGS AND FURNISHINGS					**4.A.1 Fittings, fixtures and furniture** Fixed and loose fittings and furniture including shelving, cupboards, Wardrobes, benches, seating, counters and the like. Blinds, blind boxes, curtain tracks and pelmets. Blackboards, pin-up boards, notice boards, signs, lettering, mirrors and the like. Ironmongery. **4.A.2 Soft furnishings** Curtains, loose carpets or similar soft furnishing materials. **4.A.3 Works of art** Works of art if not included in a finishes element or elsewhere. **Note:** Where items in this element have a significant effect on other elements a note should be included in the appropriate element. **4.A.4 Equipment** Non-mechanical and non-electrical equipment related to the function or need of the building (e.g. gymnasia equipment).

Fittings and furnishings	Quantity	Cost £
4.A.1 Fittings, fixtures and furniture		
4.A.2 Soft furnishings		
4.A.3 Works of art		
4.A.4 Equipment		

Preliminaries ....% of remainder of contract sum.

5.A SANITARY APPLIANCES

Fittings to be noted as grouped or dispersed.

Baths, basins, sinks, etc.
WCs, slop-sinks, urinals and the like.
Toilet-roll holders, towel rails, etc.
Traps, waste fittings, overflows and taps as appropriate.

Sanitary appliances type and quality	Number	Cost of unit £	Total £

Preliminaries ....% of remainder of contract sum.

5.B SERVICES EQUIPMENT

Kitchen, laundry, hospital and dental equipment, and other specialist mechanical and electrical equipment related to the function of the building.

Note:
Local incinerators shall be included with 'Refuse disposal' (5.C.2).

Service equipment type and quality	Number	Cost of unit £	Total £

Preliminaries ....% of remainder of contract sum.

Element and design criteria	Total cost of element £	Cost of element per m² of gross floor area £	Element unit quantity	Element unit rate £	Specification
5.C DISPOSAL INSTALLATIONS Number of sanitary appliances and special services equipment served should be stated.			–	–	**5.C.1 Internal drainage** Waste pipes to 'Sanitary appliances' (5.A) and 'Services equipment' (5.B). Soil, anti-syphonage and ventilation pipes. Rainwater downpipes. Floor channels and gratings and drains in ground within buildings up to external face of external walls. **Note:** Rainwater gutters are included in 'Roof drainage' (2.C.3). **5.C.2 Refuse disposal** Refuse ducts, waste disposal (grinding) units, chutes and bins. Local incinerators and flues thereto. Paper shredders and incinerators.

	Total £	Cost per m² of gross floor area £
Disposal installations		
5.C.1 Internal drainage		
5.C.2 Refuse disposal		

Preliminaries ....% of remainder of contract sum.

5.D WATER INSTALLATIONS				5.D.1 **Mains supply**

5.D WATER INSTALLATIONS

5.D.1 Mains supply

Incoming water main from external face of external wall at point of entry into building including valves, water metres, rising main to (but excluding) storage tanks and main taps. Insulation.

5.D.2 Cold water service

Storage tanks, pumps, pressure boosters, distribution pipework to sanitary appliances and to services equipment.

Valves and taps not included with 'Sanitary appliances' (5.A) and/or 'Services equipment' (5.B). Insulation.

Note:

Header tanks, cold water supplies, etc., for heating systems should be included in 'Heat source' (5.E).

Number of cold water draw-off points ....

5.D.3 Hot water service

Hot water and/or mixed water services.

Storage cylinders, pumps, calorifiers, instantaneous water heaters, distribution pipework to sanitary appliances and services equipment.

Valves and taps not included with 'Sanitary appliances' (5.A) and/or 'Services equipment' (5.B). Insulation.

Number of hot water draw-off points ....

5.D.4 Steam and condensate

Steam distribution and condensate return pipework to and from services equipment within the building including all valves, fittings, etc. Insulation.

Note:

Steam and condensate pipework installed in connection with space heating or the like shall be included as

Number of steam and condensate draw-off points ....

(contd.)...

Element and design criteria	Total cost of element £	Cost of element per m² of gross floor area £	Element unit quantity	Element unit rate £	Specification		
						Total £	Cost per m² of gross floor area £
5.D WATER INSTALLATIONS (contd.)					appropriate with 'Heat source' (5.E) or 'Space heating and air treatment' (5.F).		
					Water installations		
					5.D.1 Mains supply 5.D.2 Cold water service 5.D.3 Hot water service 5.D.4 Steam or condensate		
Preliminaries ....% of remainder of contract sum.							
5.E HEAT SOURCE Boiler rating in kilowatts ....kW			kW (kilowatts)		Boilers, mounting, firing equipment, pressurising equipment instrumentation and control, I.D. and F.D. fans, gantries, flues and chimneys, fuel conveyors and calorifiers. Cold and treated water supplies and tanks, fuel oil and/or gas supplies, storage tanks, etc., pipework (water or steam mains) pumps, valves and other equipment. Insulation. **Notes:** 1. Chimneys and flues which are an integral part of the structure shall be included with the appropriate structural element. 2. Local heat source shall be included with 'Local heating' (5.F.4). 3. Where more than one heat source is provided, each shall be analysed separately.		
Preliminaries ....% of remainder of contract sum.							

5.F. SPACE HEATING AND AIR TREATMENT $$\frac{Tm^2*}{\text{Gross floor area}} = \frac{\quad\quad}{\quad\quad} \begin{array}{l} m^2 \\ m^2 \end{array}$$ * Where Tm^2 is the total net area in square metres of the various compartments treated or served.	Cube of treated space as defined .....Tm^3		(i) Heating only by: **5.F.1 Water and/or steam** Heat emission units (radiators, pipes, coils, etc.) valves and fittings, instrumentation and control and distribution pipework from 'Heat source' (5.E). **5.F.2 Ducted warm air** Ductwork, grilles, fans, filters, etc., instrumentation and control. **5.F.3 Electricity** Cable heating systems, off-peak heating system, including storage radiators. **Note:** Electrically-operated heat emission units other than storage radiators should be included under 'Local heating' (5.F.4). **5.F.4 Local heating** Fireplaces (except flues), radiant heaters, small electrical or gas appliances, etc. **5.F.5 Other heating systems** (ii) Air treatment: **Notes:** 1. System described as having: 'Air treated locally' shall be deemed to include all systems where air treatment (heating or cooling) is performed either in or adjacent to the space to be treated. (contd.)...

Element and design criteria	Total cost of element £	Cost of element per m² of gross floor area £	Element unit quantity	Element unit rate £	Specification
5.F. **SPACE HEATING AND AIR TREATMENT** (contd.)					'Air treated centrally' shall be deemed to include all systems where air treatment (heating or cooling) is performed at a central point and ducted to the space being treated.

2. The combination of treatments used shall be stated, i.e.:

    Heating   Dehumidification or drying
    Cooling   Filtration
    Humidification   Pressurisation

and whether inlet extract or recirculation.

3. High velocity system shall be identified as:

    Fan coil   Induction units 2 pipe
    Dual duct   Induction units 3 pipe
    Reheat   Induction units 4 pipe
    Multi-zone   Any other system (state which).

5.F.6 **Heating with ventilation** (air treated locally)
Distribution pipework ducting, grilles, heat emission units including heating calorifiers except those which are part of 'Heat source' (5.E).
Instrumentation and control.

5.F.7 **Heating with ventilation** (air treated centrally)
All work as detailed under (5.F.6) for system where air treated centrally.

5.F.8 **Heating with cooling** (air treated locally)
All work as detailed under (5.F.6) including chilled water systems and/or cold or treated water feeds. The whole of

the costs of the cooling plant and distribution pipework to local cooling units shall be shown separately.

5.F.9 **Heating with cooling** (air treated centrally)
All work detailed under (5.F.8) for system where air treated centrally.

Note:
Where more than one system is used, design criteria specification notes and costs should be given for each.

Space heating and air treatment	Total £	Cost per m² of gross floor area £
5.F.1 Water and/or steam		
5.F.2 Ducted warm air		
5.F.3 Electricity		
5.F.4 Local heating		
5.F.5 Other heating systems		
5.F.6 Heating with ventilation (air treated locally)		
5.F.7 Heating with ventilation (air treated centrally)		
5.F.8 Heating with cooling (air treated locally)		
5.F.9 Heating with cooling (air treated centrally)		

Preliminaries ....%
of remainder of contract sum.

Element and design criteria	Total cost of element £	Cost of element per m² of gross floor area £	Element unit quantity	Element unit rate £	Specification
5.G VENTILATING SYSTEM $$\frac{Tm^2*}{\text{Gross floor area}} = \underline{\quad m^2 \quad} = \underline{}$$ * Where Tm^2 is the total net area in square metres of the various compartments treated or served. Preliminaries ....% of remainder of contract sum.			Cube of treated space as defined ......Tm^3		Mechanical ventilating system not incorporating heating or cooling installations including dust and fume extraction and fresh air injection, extract fans, rotating ventilators and instrumentation and controls.
5.H. ELECTRICAL INSTALLATIONS Total electric load in kilowatts ....kW (tabulate illumination levels by principle functions)					**5.H.1 Electric source and mains** All work from external face of building up to and including local distribution boards including main switchgear, main and sub-main cables, control gear, power factor correction equipment, stand-by equipment, earthing, etc. **Notes:** 1. Installations for electric heating ('built-in' systems) shall be included with 'Space heating and air treatment' (5.F.3). 2. The cost of stand-by equipment shall be stated separately.

Electric power supplies	Total £	Cost per m² of gross floor area £
5.H.2 Electric power supplies All wiring, cables, conduits, switches from local distribution boards, etc. to and including outlet points for the following:		
Electric power supplies		
General purpose socket outlets Services equipment Disposal installations Water installations Heat source Space heating and air treatment Gas installation Lift and conveyor installations Protective installations Communication installations Special installations		
Total number of power outlets	—	
Note: The cost pf the power supply to these installations should, where possible, be shown separately. **5.H.3 Electric lighting** All wiring, cables, conduits, switches, etc., from local distribution boards and fittings to and including outlet points.		
Total number of lighting points	—	

(contd.)...

Element and design criteria	Total cost of element £	Cost of element per m² of gross floor area £	Element unit quantity	Element unit rate £	Specification
5.H. ELECTRICAL INSTALLATIONS (contd.)					**5.H.4 Electric lighting fittings** Lighting fittings including fixing. Where lighting fittings supplied direct by client, this should be stated.
Illumination in lux lx					Light fittings type and quality — Number — Cost £
					Electrical installations — Total £ — Cost per m² of gross floor area £ 5.H.1 Electric source and mains 5.H.2 Electric power supplies 5.H.3 Electric lighting 5.H.4 Electric lighting fittings
Preliminaries ....% of remainder of contract sum					
5.I GAS INSTALLATIONS Number of draw-off points			Number of draw-off points		Town and natural gas services from metre or from point of entry where there is no individual meter: distribution pipework to appliances and equipment.
Preliminaries ....% of remainder of contract sum.					

5J LIFT AND CONVEYOR INSTALLATIONS

The number of rush period passengers for which the installation has been designed should be stated.

Lifts

The number, capacity, speed, number of stops, number of doors and height served should be stated. Capacity of hoists to be given in kilogrammes.

Escalators

Rise and travel of escalators should be stated.

Preliminaries ....% of remainder of contract sum.

5.J.1 Lifts and hoists

The complete installation including gantries, trolleys, blocks, hooks and ropes, downshop leads, pendant controls and electrical work from and including isolator.

Notes:
1. The cost of special structural work, e.g. lift walls, lift motor rooms, etc. shall be included in the appropriate structural elements.
2. Remaining electrical work shall be included with 'Electric power supplies' (5.H.2).
3. The cost of each type of lift or hoist shall be stated separately.

5.J.2 Escalators
As detailed under 5.J.1

5.J.3 Conveyors
As detailed under 5.J.1

	Total £	Cost per m² of gross floor area £
Life and conveyor installations		
5.J.1 Lifts and hoists		
5.J.2 Escalators		
5.J.3 Conveyors		

Element and design criteria	Total cost of element £	Cost of element per m² of gross floor area £	Element unit quantity	Element unit rate £	Specification
5.K PROTECTIVE INSTALLATIONS Sprinklers – the number of outlets, control mechanism and area served by each control mechanism should be given. $$\frac{Tm^2*}{\text{Gross floor area}} = \underline{\hspace{2cm}} \; m^2$$ $$= \underline{\hspace{2cm}} \; m^2$$ * Where Tm² is the total net area in square metres of the various compartments treated or served.					**5.K.1 Sprinkler installations** The complete sprinkler installation and CO_2 extinguishing system including tanks control mechanism, etc. **Note:** Electrical work shall be included with 'Electric power supplies' (5.H.2). **5.K.2 Fire-fighting installations** Hosereels, hand extinguishers, asbestos blankets, water and sand buckets, foam inlets, dry risers (and wet risers where only serving fire fighting equipment). **5.K.3 Lightning protection** The complete lightning protection installation foam finials conductor tapes, to and including earthing. **Note:** The cost of lightning protection to boiler and vent stacks shall be stated separately.

	Total £	Cost per m² of gross floor area £
Protective installations		
5.K.1 Sprinkler installations 5.K.2 Fire-fighting installations 5.K.3 Lightning protection		

Preliminaries ....%
of remainder of contract sum.

5.L COMMUNICATION INSTALLATIONS

The following installations shall be included:

Communication installations	Total £	Cost per m² of gross floor area £
Warning installations (fire and theft)		
Burglar and security alarms		
Fire alarms		
Visual and audio installations		
Door signals		
Timed signals		
Call signals		
Clocks		
Telephones		
Public address		
Radio		
Television		
Pneumatic message systems		

Notes:

1. The cost of each installation shall be stated separately if possible along with an indication of the specification.

2. The cost of the work in connection with electrical supply shall be included with 'Electric power supplies' (5.H.2).

Preliminaries ....% of remainder of contract sum.

Element and design criteria	Total cost of element £	Cost of element per m² of gross floor area £	Element unit quantity	Element unit rate £	Specification
5.M SPECIAL INSTALLATIONS					All other mechanical and/or electrical installations (separately identifiable) which have not been included elsewhere, e.g. Chemical gases; Medical gases; Vacuum cleaning; Window cleaning equipment and cradles; Compressed air; Treated water; Refrigerated stores. **Notes:** 1. The cost of each installation shall, where possible, be shown separately along with an indication of the specification. 2. Items deemed to be included under 'Refrigerated stores' comprise all plant required to provide refrigerated conditions (i.e. cooling towers, compressors, instrumentation and controls, cold room thermal insulation and vapour sealing, cold room doors, etc.) for cold rooms, refrigerated stores and the like other than that required for 'Space heating and air treatment' (5.F.8 and 5.F.9).
Preliminaries ....% of remainder of contract sum.					
5.N BUILDER'S WORK IN CONNECTION WITH SERVICES			–	–	Builder's work in connection with mechanical and electrical services. **Notes:** 1. The cost of builder's work in connection with each of the services elements shall, where possible, be shown separately.

2. Where tank rooms, housings and the like are included in the gross floor area, their component parts shall be analysed in detail under the appropriate elements. Where this is not the case the cost of such items shall be included here.

Services elements	£	Cost of BWIC per m² of gross floor area £
5.A Sanitary appliances		
5.B Services equipment		
5.C Disposal installations		
5.D Water installations		
5.E Heat source		
5.F Space heating and air treatment		
5.G Ventilating system		
5.H Electrical installations		
5.I Gas installations		
5.J Lift and conveyor installations		
5.K Protective installations		
5.L Communication installations		
5.L Special installations		

Preliminaries ....% of remainder of contract sum

Element and design criteria	Total cost of element £	Cost of element per m² of gross floor area £	Element unit quantity	Element unit rate £	Specification		
5.O **BUILDER'S PROFIT AND ATTENDANCE ON SERVICES**					Builder's profit and attendance in connection with mechanical and electrical services. **Note:** The cost of profit and attendance in connection with each of the services elements shall, where possible, be shown separately.		
							Cost of BA & A per m² of gross floor area £
					Services elements	£	
					5.A Sanitary appliances 5.B Services equipment 5.C Disposal installations 5.D Water installations 5.E Heat source 5.F Space heating and air treatment 5.G Ventilating system 5.H Electrical installations 5.I Gas installations 5.J Lift and conveyor installations 5.K Protective installations 5.L Communication installations 5.L Special installations		
Preliminaries .... % of remainder of contract sum.							
Sub-total excluding External works, Preliminaries and Contingencies £							

Element and design criteria	Total cost of element £	Specification	Cost of sub-element £	Element unit quantity	Element unit rate £
6.A SITE WORKS Cost per m² gross floor area £..........		**6.A.1 Site preparation** Clearance and demolitions Preparatory earth works to form new contours		) m²)	
		6.A.2 Surface treatment The cost of the following items shall be stated separately if possible: Roads and associated footways Vehicle parks Paths and paved areas Playing fields Playgrounds Games courts Retaining walls Land drainage Landscape work		))))) m²))))	
		6.A.3 Site enclosure and division Gates and entrance Fencing, walling and hedges		) m)	
		6.A.4 Fittings and furniture Notice boards, flag poles, seats, signs		number	
Preliminaries ....% of remainder of contract sum.				–	–
6.B DRAINAGE Cost per m² gross floor area £.		Surface water drainage Foul drainage Sewage treatment. **Note:** To include all drainage works (other than land drainage included with 'Surface treatment' 6.A.2)) outside the building to and including disposal point, connection to sewer or to treatment plant.			
Preliminaries ....% of remainder of contract sum.					

Element and design criteria	Total cost of element £	Specification	Cost of sub-element £	Element unit quantity	Element unit rate £
6.C EXTERNAL SERVICES Cost per m² gross floor area £..................		**6.C.1 Water mains** Main from existing supply up to external face of building. **6.C.2 Fire mains** Main from existing supply up to external face of building; fire hydrants **6.C.3 Heating mains** Main from existing supply or heat source up to external face of building. **6.C.4 Gas mains** Main from existing supply up to external face of building. **6.C.5 Electric mains** Main from existing supply up to external face of building. **6.C.6 Site lighting** Distribution, fittings and equipment. **6.C.7 Other mains and services** Mains relating to other service installations (each shown separately). **6.C.8 Builder's work in connection with external services** Builder's work in connection with external mechanical and electricalservices: e.g. pits, trenches, ducts, etc.			
		6.C.1		–	–
		6.C.2		–	–
		6.C.3		–	–
		6.C.4		–	–
		6.C.5		–	–
		6.C.6		–	–
		6.C.7		–	–

6.C EXTERNAL SERVICES (contd.)	**Note:**
	The cost of the builder's work shall be stated separately for each of the sub-sections (6.C.1) to (6.C.7).
	
	6.C.9 **Builder's profit and attendance on external mechanical and electrical services**
	
	      6.C.1
	      6.C.2
	      6.C.3
	      6.C.4
	      6.C.5
	      6.C.6
	      6.C.7
	
	Note:
	The cost of profit and attendances shall be stated separately for each of the sub-sections (6.C.1) to (6.C.7).
			–	–

Preliminaries ....% of remainder of contract sum.				
6.D MINOR BUILDING WORK Cost per m² gross floor area £................	6.D.1 **Ancillary buildings** Separate minor buildings such as sub-stations, bicycle stores, horticultural buildings and the like, inclusive of local engineering services. 6.D.2 **Alterations to existing buildings** Alterations and minor additions, shoring, repair and maintenance to existing buildings.	Gross floor area of ancillary buildings (m²) –	–	
Preliminaries ....% of remainder of contract sum.				
PRELIMINARIES Cost per m² gross floor area £................ ....% of remainder of contract sum.	Priced items in Preliminaries Bill and Summary but excluding contractor's price adjustments. Individual costs of the main preliminary items should be given. **Notes:** (i) Professional fees will not form part of the cost analysis. (ii) Lump sum adjustments shall be spread pro-rata amongst all elements of the building and external works based on all work excluding Prime Cost and Provisional Sums.			
TOTAL (less contingencies)				

Drawings: Drawings should always accompany an Amplified Cost Analysis.

5: LETTER SEEKING PERMISSION TO SUBMIT COST ANALYSIS TO BCIS

Where cost analyses are submitted by subscribers to BCIS, it is understood that permission has been obtained from the appropriate authority and the following letters have been drafted to facilitate this.

Suggested form of letter to the Building Owner and the Contractor

Dear Sir

Name of Project

We are subscribers to the Royal Institution of Chartered Surveyors' Building Cost Information Service, and are collaborating in the production of statistics concerning costs of buildings. The above is one of the projects in respect of which we would wish to report cost information.

Costs will be reported under element headings but individual bill rates will not be quoted. The names of the building owner, the architect, quantity surveyor, consulting engineers and general contractor will normally be given. However, should you so require it, BCIS will not publish these names nor identify the precise name and location of the project. This information will be kept confidential to subscribers of the Building Cost Information Service. Subject to the architect's and your own approval, sketch plans and elevations will be included to illustrate the design/shape of the block.

As the information contained in priced Bills of Quantities is confidential to the contracting parties, we do not feel that we could properly make use of it even in this form of cost analysis without the permission of yourselves and the contractor (or the building owner).

We should be grateful if you would kindly give your permission in this case. If possible, it would be most helpful if you would consent to give blanket approval, so far as your own interest is concerned, to any future information of this kind being reported by us in respect of contracts with which we are jointly concerned without the need of further specific requests in each case.

We are similarly asking the permission of the contractor (building owner).

Yours faithfully,

Suggested form of letter to the Architect

Dear Sir

Name of Project

We are subscribers to the Royal Institution of Chartered Surveyors' Building Cost Information Service, and are collaborating in the production of statistics concerning tender costs of building. The above is one of the projects in respect of which we would wish to submit a cost analysis.

It is felt that if sketch plans and elevations accompanied the analysis they would help the members of BCIS to interpret the costs with a better knowledge of the contract. The drawings would convey the design and shape of the building far better than any wordy description could do. The purpose of this letter is, therefore, to ask your permission to submit copies of sketch plans and elevations to be reproduced along with the analysis.

We should be grateful if you would kindly give your permission in this case. If possible, it would be most helpful if you would consent to allow us to submit drawings in respect of other contracts with which we are jointly concerned without the need of further specific requests in each case.

Yours faithfully,

Index